Tourism and Postcolonialism

Tourism studies has the potential to make a significant contribution to understanding the postcolonial experience because of the centrality of tourism to the processes of transnational mobilities, migration and globalisation.

Tourism and Postcolonialism draws together theoretical and applied research in order to illuminate the links between tourism, colonialism and postcolonialism. Significantly, the book aims to create a space for the voices of authors from postcolonial countries. Chapters are integrated and examined through concepts drawn from the wide postcolonial literature, and identify tourism not only as a postcolonial cultural form, but as an international industry which is based on past and present-day colonial structural relationships.

C. Michael Hall is Professor of Tourism, and **Hazel Tucker** is Senior Lecturer of Tourism, at the Department of Tourism, University of Otago, New Zealand.

Contemporary Geographies of Leisure, Tourism and Mobility

Series editor: C. Michael Hall, Professor of Tourism, University of Otago, New Zealand.

The aim of this series is to explore and communicate the intersections and relationships between leisure, tourism and human mobility within social sciences.

It will incorporate both traditional and new perspectives on leisure and tourism from contemporary geography, e.g. notions of identity, representation and culture, while also providing for perspectives from cognate areas such as anthropology, cultural studies, gastronomy and food studies, marketing, policy studies and political economy, regional and urban planning, and sociology, within the development of an integrated field of leisure and tourism studies.

Also, increasingly, tourism and leisure are regarded as steps in a continuum of human mobility. Inclusion of mobility in the series offers the prospect to examine the relationship between tourism and migration, the sojourner, educational travel, and second home and retirement travel phenomena.

The series comprises two strands:

Contemporary Geographies of Leisure, Tourism and Mobility aims to address the needs of students and academics, and the titles will be published in hardback and paperback. Titles include:

The Moralisation of Tourism
Sun, sand . . . and saving the world?
Jim Butcher

The Ethics of Tourism Development
Mick Smith and Rosaleen Duffy

Tourism in the Caribbean
Trends, development, prospects
Edited by David Timothy Duval

Qualitative Research in Tourism
Ontologies, epistemologies and methodologies
Edited by Jenny Phillimore and Lisa Goodson

Routledge Studies in Contemporary Geographies of Leisure, Tourism and Mobility is a forum for innovative new research intended for research students and academics, and the titles will be available in hardback only. Titles include:

1. Living with Tourism
Negotiating identities in a Turkish village
Hazel Tucker

2. Tourism, Diaspora and Space
Tim Coles and Dallen J. Timothy

3. Tourism and Postcolonialism
Contested discourses, identities and representations
C. Michael Hall and Hazel Tucker

Tourism and Postcolonialism

Contested discourses, identities
and representations

**C. Michael Hall and
Hazel Tucker**

Routledge
Taylor & Francis Group

LONDON AND NEW YORK

First published 2004
by Routledge
2 Park Square, Milton Park, Abingdon, Oxon OX14 4RN

Simultaneously published in the USA and Canada
by Routledge
270 Madison Ave, New York, NY 10016

Transferred To Digital Printing 2006

Routledge is an imprint of the Taylor & Francis Group

Typeset in Times by
Florence Production Ltd, Stoodleigh, Devon
Printed and bound in Great Britain
by MPG Digital Solutions, Bodmin, Cornwall

British Library Cataloguing in Publication Data
A catalogue record for this book is available from the British Library

Library of Congress Cataloging in Publication Data
A catalog record for this book has been requested

ISBN 0–415–33102–1

To the Wandering Islands
(and Blobby)

Contents

Contributors

John S. Akama Department of Tourism, Moi University, Kenya.

Hilary du Cros Department of Hotel and Tourism Management, Hong Kong Polytechnic University, Kowloon, Hong Kong, SAR, China.

David Timothy Duval Department of Tourism, University of Otago, New Zealand.

David Fisher Department of Human Sciences, Lincoln University, New Zealand.

C. Michael Hall Department of Tourism, University of Otago, New Zealand.

Joan C. Henderson Nanyang Business School, Nanyang Technological University, Singapore.

Keith Hollinshead Luton Business School, The University of Luton, England.

Reiner Jaakson Department of Geography, University of Toronto, Ontario, Canada.

Sabine Marschall University of Durban-Westville, Durban, South Africa.

Beverley Ann Simmons Department of Sociology and Anthropology, The University of Newcastle, New South Wales, Australia.

Hazel Tucker Department of Tourism, University of Otago, New Zealand.

Harry Wels Department of Culture, Organisation and Management, Vrije Universiteit Amsterdam, Amsterdam, The Netherlands.

Preface

Although being an area of great intellectual richness, postcolonialism is also an area of contestation and confusion. The range of contributions to this volume provides good evidence of that. As two academics of European heritage living in a land with its own colonial past and postcolonial present, we are also acutely aware of the tensions involved in interrogating postcolonialism not only in its academic form but also in relation to its day-to-day realities. In undertaking the work that led to this volume, we are therefore conscious that this book seeks to provide a space for different voices to be heard on a topic that, for whatever reasons, has been ignored from much study of tourism, even in postcolonial societies themselves. However, we are also extremely aware that postcolonial pedagogy and research itself needs to be understood within the context of institutional circuits of production and consumption in which it has substantial commodity status. As Bahri observes:

> The contradictions inherent in the institutionalization of difference pose a persistent challenge to those who seek to remain critical of the very system that has accorded them their authority and their position. . . . As teachers drawn in many cases from the elite ranks of universities in ex-colonies, our dilemma is compounded because some of us both teach and embody the margins. We teach, 'translate,' and make available through a filter of postcolonial history and theory the 'voices' (nothing less than the 'voice' will do, given our rhetoric of speaking and being listened to as if an actual exchange were being enacted that transcended the merely academic) simultaneously reinstated in the periphery as they are introduced into the discourse at the center.
>
> (Bahri 1997: 279)

To note Bahri and her critique of much postcolonial writing and theory is therefore to reinforce the notion that critical intervention through an examination of postcolonial pedagogy and theory must be formulated within a thorough understanding of its institutional and discursive context and the power relations of the academy. Therefore, we are more than aware of the limitations of discourse on postcolonialism and the issue of the voices that are heard. However, such a situation should not stop conversations being initiated. Indeed, the opportunity to give voice to postcolonial concerns was not taken by all potential contributors. That said, we hope that this book will represent a significant step in bringing postcolonialism and tourism studies closer together for the mutual benefit that a discourse between the two fields may provide as well as the light that may be shed on core issues of heritage, representation and identity.

In completing this book, we would like to thank Monica Gilmour, Peter Treloar and our other colleagues in the Department of Tourism, University of Otago, for their support. Andrew Mould and Melanie Attridge of Routledge also provided enormous support for the project and demonstrated great patience with the editors when the manuscript was unexpectedly delayed. Finally, we would like to proffer our personal thanks. Hazel would like to extend her thanks to family and friends for their support, particularly when the book was being completed, while Michael would like to do the same, particularly noting the contributions of David Duval, Tim Coles and Allan Williams to thinking about the relations between migration and postcolonialism, as well as thanking Jody for coping with yet another Christmas book.

We would like to conclude this Preface by noting that any discourse is imperfect, perhaps especially so in a globalised, postcolonial world. Yet to encourage critical discourse and communication in the tourism academy at a time when some governments seem to be abandoning multilateralism and embarking on new neocolonial adventures seems to be the most appropriate course of all.

C. Michael Hall and Hazel Tucker
Dunedin

Reference

Bahri, D. (1997) 'Marginally off-center: postcolonialism in the teaching machine', *College English* 59, 3: 277–98.

1 Tourism and postcolonialism

An introduction

C. Michael Hall and Hazel Tucker

Postcoloniality arguably became the central, controversial site for literary studies in the last decade of the twentieth century, or what could claim to have been, more than anything else, the imperial (and the colonial) century.

The 'postcolonial' appears to signify challenge yet, of course, literary challenges to the hegemonic power of the centre are not new phenomena. But, as the authors of *The Empire Writes Back* (Ashcroft *et al.* 1989) have demonstrated, there is something particularly potent (something powerfully challenging) about the current set of so-called 'postcolonial texts'. While acknowledging the potency of much recent writing in this field, it is true to say that the central question – what constitutes a postcolonial text – remains a contentious issue. If we follow Edward Said's thought that 'to be one of the colonized is potentially to be a great many different, but inferior, things, in many different places, at many different times', there is no reason to think that to be one of the *post* colonised is a homogenous position (de Reuck and Webb 1992).

The concept of postcolonialism, which for much of the 1990s has informed cultural theorising, is increasingly influencing the intellectual terrain of tourism studies. Studies of tourism in the less developed countries, concerns over identity and representation, and theorising over the nature and implications of the cultural, political and economic encounters that are intrinsic to the tourist experience, have increasingly led to reference to postcolonial discourse. However, such examination of and reference to intellectual space should not be seen as occurring in an uncritical fashion. Instead, postcolonial analysis in tourism reflects the essential contested nature of postcolonial studies elsewhere in the social sciences and humanities. Indeed, the oft-noted difficulty of finding an acceptable definition and academic ground with which to describe tourism studies is no different from the experiences of those engaged in postcolonial studies. Nevertheless, just because something is hard to describe does not mean it is not there or that it is unimportant.

The purpose of this book is to examine some of the various means by which the idea of the postcolonial may contribute to tourism studies. In this it seeks to present not only the means by which postcolonial thought may be of relevance to the study of tourism, but also how tourism itself may shed some insights on the postcolonial. However, it is remarkable that recent key texts in the postcolonial field (e.g. Loomba 1998; Young 2001; Goldberg and Quayson 2002) have failed to acknowledge the potential contribution that tourism studies can make to understanding the postcolonial experience (Edensor 1998), despite the centrality of tourism to the processes of transnational mobilities and migrations, and globalisation (Hall and Williams 2002; Coles and Timothy 1994; Coles *et al.* 2004; Hollinshead, this volume). As Craik (1994) recognised:

> Tourism has an intimate relationship to post-colonialism in that ex-colonies have increased in popularity as favoured destinations (sites) for tourists (the Pacific Rim; Asia; Africa; South America); while the detritus of post-colonialism have been transformed into tourist sights (including exotic peoples and customs; artefacts; arts and crafts; indigenous and colonial lifestyles, heritage and histories).
>
> (Ibid.)

Tourism therefore both reinforces and is embedded in postcolonial relationships. Issues of identity, contestation and representation are increasingly recognised as central to the nature of tourism, particularly given recent reflection on the ethical bases of tourism and tourism studies (Butcher 2003). However, much of this discussion has tended to take place on what are, arguably, the fringes of academic tourism discourse, although such issues have received more attention in cultural geography, anthropology and cultural studies.

Postcolonialism represents both a reflexive body of Western thought that seeks to reconsider and interrogate the terms by which the duality of coloniser and colonised, with its accompanying structures of knowledge and power, has been established as well as the state of being 'post' or 'after' the condition of being a colony. Students of postcolonialism are therefore interested in spatial and temporal dimensions of the cultural production and social formation of the colony and postcolony and the ongoing construction and representation of specific spaces and experiences. Examination of neocolonial relationships, a situation in which an independent country continues to suffer intervention and control from a foreign state, is also often incorporated into the postcolonial corpus, although in recent years, rather than just refer to external state intervention, neo-colonialism has also been used to refer to the expansion of capitalism and economic and cultural globalisation so that the core powers exercise influence over the postcolonial periphery. In many of these cases the term postcolonial is often applied to jurisdictions that have yet to achieve political independence but which remain highly peripheral. In addition, the term is also applied to internal spatial and social peripheries, including minorities that are dominated by a metropolitan core. More generically, the postcolonial can refer to a position against imperialism, colonialism and Eurocentrism, including Western thought and philosophy. However, as noted below, such a situation may create numerous tensions and contradictions that have not been resolved within postcolonial studies.

Although there are exceptions (for example, see the discussion in Finnström 1997), the relationship between (post)coloniser and (post)-colonised is primarily seen in the context of the interactions between European nations and the regions and societies they colonised since the onset of European mercantile expansion and imperialism in the fifteenth century, and its subsequent disintegration. Undoubtedly, the European imperial influence was considerable, incorporating over 80 per cent of the earth's land surface by the commencement of the First World War in 1914 (interestingly, much of what remained outside the European powers' imperial domains was controlled by the colonial powers of Japan and the United States). However, it is important to note that the state of being postcolonial, particularly in tourism, may not be directly informed by what is generally referred to as postcolonial studies. Both of these representations of postcolonialism are to be found in the contributions to the present volume. This first chapter seeks to introduce some of the concerns and issues within postcolonial studies and indicate their relevance to contemporary studies of tourism, as well as wider concerns with colonial relationships.

Positioning postcolonialism

As a terrain of knowledge the concept of postcolonialism is problematic and at once contested (Bahri 1995, 1997). Indeed, a concept such as postcolonial never ends a discussion, it begins it. Labelling something, such as an event or text, or even an attitude, as 'postcolonial' therefore places it within a broad category of things under discussion. Such postcolonial sites of argument and questioning encompass different scales, from the local to the global, and incorporate issues of geographic, cartographic, cultural, economic, gender,· literary, political and socio-linguistic specificity and heterogeneity. Although the development of postcolonial studies has been heavily influenced by Said's seminal work on *Orientalism* (1978) and the development of the notion of the other in Western thought, arguably one of the lynchpins of postcolonial thought was Ashcroft *et al.*'s study of postcolonial literature, *The Empire Writes Back* (1989). Ashcroft *et al.* (1989: 2) used the term 'postcolonial' (also 'post-colonial') 'to cover all the culture affected by the imperial process from the moment of colonization to the present day. This is because there is a continuity of preoccupations throughout the historical process initiated by European imperial aggression'. They also suggested that as a term it was the most appropriate to describe 'the new cross-cultural criticism which has emerged in recent years and for the discourse through which this is constituted' (1989: 2). In the case of the latter, in literary and cultural studies cognate terms such as

'Commonwealth' (with references to the countries and literatures of the former British empire and members of the present-day Commonwealth) and 'Third World', which were used to describe the literature of Europe's former colonies, have certainly become rarer and have tended to be replaced by the term 'postcolonial'. However, interestingly, in tourism this term has not been so readily adopted and instead the notion of 'developing' or 'less developed' countries has been far more significant in replacing the concept of 'Third World' (e.g. Harrison 1992, 2001), possibly because of the historically greater influence of theories of economic development on the field than literary studies.

Ashcroft *et al.* (1989) identified four main areas of interrelated investigation in postcolonial studies which continue to inform the postcolonial project to the present day: hegemony, language and text, place and displacement and the development of theory, and it is to these we will now turn.

Hegemony

Ashcroft *et al.* (1989) posed the question as to why postcolonial societies should continue to engage with the imperial experience, since nearly all postcolonial societies have achieved political independence. Why is the issue of coloniality still relevant at all? Here debate is substantially focused on the ongoing political, economic and cultural influence of the former imperial powers, often regarded as including the United States, in postcolonial states as well as the deep inequalities that exist between North and South (Ferro 1997). Much of this debate has focused on the core-periphery relationship that exists in economic and political terms between the developed and the less-developed countries, as well as some debates on internal peripheries, and this has had some influence on the tourism literature, particularly in the 1970s and 1980s. For example, Matthews described tourism as potentially being a new colonial plantation economy in which 'Metropolitan capitalistic countries try to dominate the foreign tourism market, especially in those areas where their own citizens travel most frequently' (1978: 79). Air services, bus companies, hotels, resort developments, recreational facilities such as golf courses and food and beverage are all potential markets related directly to tourism which may become owned by foreign interests (see Jaakson, this volume). The elements of a plantation tourism economy are that:

1   tourism is structurally a part of an overseas economy;
2   it is held together by law and order directed by the local elites;
3   there is little or no way to calculate the flow of values.

(Best 1968)

Matthews' thesis was developed in relation to the influence of American and multinational corporations on Caribbean tourism development, but may well be applied to other situations where core-periphery relations are seen to exist, particularly with respect to island microstates. In the case of the Pacific, it has been argued that tourism development, along with other foreign economic services such as tax havens, demonstrates elements of a plantation economy (Britton 1982a, 1982b, 1983; Connell 1988), in which the island nations are nothing more than the place of production in a system of trade and production in which control lies with the demand for produce in the First World and with the merchants (Girvan 1973). Within the plantation economy overseas interests are critical for creating both the demand and supply of the tourist product. For example, Britton argued:

> without the involvement of foreign and commercial interests, Tonga has not evolved the essential ties with metropolitan markets and their tourism companies. It would seem that Tonga's tourist industry has paradoxically suffered because the country was not exploited as a fully-fledged colony.
>
> (1987: 131)

Clearly, such a situation also reflects one of the ironies of postcoloniality that in terms of the development of international economic networks then being a former colony may be advantageous. However, as Matthews cautioned:

> Tourism may add to the numbers of jobs available and it may increase the trappings of modernity with modern buildings and new services, but if it does not contribute to the development of local resources, then it differs little from the traditional agricultural plantation.
>
> (1978: 80)

The situation of economic and political dependency arising out of sets of postcolonial core-periphery relationships has been likened by some commentators to a form of imperialism. For example, Crick argues that tourism is a form of 'leisure imperialism' and represents 'the hedonistic face of neocolonialism' (1989: 322). Similarly, Nash perceived the concept quite broadly:

> At the most general level, theories of imperialism refer to the expansion of a society's interests abroad. These interests – whether economic, political, military, religious, or some other – are imposed on or adopted by an alien society, and evolving intersocietal transactions, marked by the ebb and flow of power, are established.
>
> (1989: 38).

According to Nash:

> Metropolitan centers have varying degrees of control over the nature of tourism and its development, but they exercise it – at least at the

beginning of their relationship with tourist areas – in alien regions. It is this power over touristic and related developments abroad that makes a metropolitan center imperialistic and tourism a form of imperialism.

(1989: 39)

However, the extent to which power is able to be exercised, and hence development is controlled in any nation or destination by an external agency is somewhat problematic as a more complex notion of globalisation has replaced simplistic ideas of imperialism (Hall 1998). Indeed, there is a general failure of critics of cultural imperialism to grasp fully the ambiguous gift of capitalist modernity inherent in contemporary globalisation, that is, there is a need to probe the contradictions of capitalist culture and its implications for tourism (Britton 1991; Jaakson, this volume). Nevertheless, in terms of the relationships between the former colonisers and the colonised it is apparent that a substantial legacy continues to exist with respect to political economy that clearly may have relevance for the pattern and nature of tourism development and, of course, for the wider society. This observation continues to have resonance in some more recent analyses of tourism (e.g. Mowforth and Munt 1998; Meethan 2001), but arguably the condition of postcoloniality and the power relationships that it situates have not received anywhere near the level of overt recognition or interrogation in tourism studies that it deserves.

Language, text and representation

According to Ashcroft *et al.* (1989) one of the main features of imperial oppression is control over language and text. This occurred because the imperial education system installed a 'standard' version of the metropolitan language, e.g. the notion of the Queen's English, as the norm, and other versions as impurities. Furthermore, they note that language becomes the medium through which a hierarchical structure of power is perpetuated, and the medium through which conceptions of 'truth', 'order' and 'reality' become established. Therefore, much of the discussion of postcolonial writing in literary and cultural studies is focused on how language and writing, with its power and signification of authority can be wrested from the dominant European culture in order to provide an effective postcolonial voice (Bhabha 1984; Ashcroft *at al.* 1989). For example, Batsleer *et al.* (1985) observed that the historical moment which saw the emergence of 'English' as an academic discipline also produced the nineteenth-century colonial form of imperialism. Ashcroft *et al.* argue:

that the study of English and the growth of Empire proceeded from a single ideological climate and that the development of the one is intrinsically bound up with the development of the other, both at the level of simple utility (as propaganda for instance) and at the unconscious level, where it leads to the naturalizing of constructed values (e.g. civilization, humanity, etc.) which, conversely, established 'savagery', 'native', 'primitive', as their antitheses and as the object of a reforming zeal.

(1989: 4)

Nevertheless, English was not the only discipline developed in relation to the imperial mission. The field of geography also owes much to European mercantile and imperialist expansion (Johnston 1991; Livingstone 1992; Hall and Page 2002). For example, the first issue of the British Royal Geographical Society journal made plain the raison d'être of the new scientific body:

> That a new and useful society might be formed, under the name of THE ROYAL GEOGRAPHICAL SOCIETY OF LONDON.
> That the interest excited by this department of science is universally felt; that its advantages are of the first importance to mankind in general, and paramount to the welfare of a maritime nation like Great Britain, with its numerous and extensive foreign possessions.
> (*Journal of the Royal Geographical Society of London*,
> 1 (1831): vii; in Livingstone 1992: 167)

As Livingstone observed, 'This submission makes plain the imperialistic undergirding of the institution's entire project and thereby reveals that Victorian geography was intimately bound up with British expansionist policy overseas' (1992: 167). However, while the development of other disciplines, such as anthropology (Ranger 1976), botany (Brockway 1979) and geology (Stafford 1984; MacKenzie 1990), was also bound up with overseas expansion:

> there is something to be said that geography was the science of imperialism par excellence. Exploration, topographic and social survey, cartographic representation, and regional inventory – the craft practices of the emerging geographical professional – were entirely suited to the colonial project.

(Livingstone 1992: 170)

While geography was an imperialist means of control it also played a vital role in establishing representations of non-European others in the imperial mind. Within its 'scientific' method, geography was able to establish and represent stereotypes of race, ethnicity, economy and culture that exist to the present day but which are being critically reevaluated in contemporary geographies of postcolonialism (e.g. Barnett 1997, 1998; Blunt and McEwan 2002; Clayton 2003). Historically, the geographical project of nineteenth-century imperialism did much to create the idea of orientalism.

8   *C. Michael Hall and Hazel Tucker*

Orientalism is a style of thought based upon ontological and epistemological distinction made between 'the Orient' and (most of the time) 'the Occident'. From the mid-nineteenth century to the present day, very many writers have accepted the basic distinction between East and West as the 'starting point for elaborate accounts concerning the Orient, its people, customs, "mind", destiny, and so on . . . despite or beyond any correspondence, or lack thereof, with a "real" Orient' (Said 1978: 5). Such otherness is essential in tourism (Hall 1998). 'Encounters with the "other" have always provided fuel for myths and mythical language. Contemporary tourism has developed its own promotional lexicon and repertoire of myths . . .' (Selwyn 1993: 136). For the vast majority of people, otherness is what makes a destination worthy of consumption. Although, ironically, 'large numbers of tourists may be attracted to the region by its perceived "differentness", lured by the images of culture and landscape which are vividly portrayed in the promotional literature, few are able or willing to tolerate a great deal of novelty' (Hitchcock *et al.* 1993: 3). However, to build binary opposites is to make one dependent on the other. There cannot be consumption without production. 'It is apparent that they merge in many places and that each process certainly does have effects on the other . . . even if they are causal or may never ever be explicable' (Laurier 1993: 272). Any understanding of the creation of a destination therefore involves placing the development of the representation of that destination within the context of the historical consumption and production of places and the means by which places have become incorporated within the global capital system. Moreover, such an analysis leads to the recognition that the post-colonial experience is also related to the subjugation and utilisation of nature for the colonial powers. In this sense it is possible to talk of eco-colonialism (Mowforth and Munt 1998) and eco-imperialism (Hall 1994b) not only in an historical setting, but also with respect to present-day ecotourism (Akama, this volume). As Hall concluded:

In his book *Ecological Imperialism: The Biological Expansion of Europe, 900–1900*, Crosby (1986) describes the, sometimes forced, Europeanisation of the global environment through the spread of the plant and animal species most desired by the European peoples. In the current age of supposed environmental awareness, many European peoples are seeking ways to restrain gene, species, and ecosystem loss and preserve biodiversity through national park and associated reserve systems. Ecotourism is being promoted throughout the world . . . as a means to achieve both environmental conservation and economic return in conjunction with these systems. Undoubtedly, the maintenance of biodiversity is a critical component of sustainable development. However, sustainable development also teaches us that environment and economy are integrated with society. Many promoters of ecotourism . . . have either forgotten or ignored this lesson. Therefore, we are perhaps facing a form of ecological imperialism in the region in which a new set of European cultural values are being impressed on indigenous cultures through ecotourism development.

(1994b: 153–4)

The representation of otherness was, and still is, also inextricably linked to the popularisation of accounts of travels and explorations in the imperial lands (Foster 1990; Pratt 1992; Spurr 1993; Hall 1998; Clark 1999; Glage 2000; Simmons, this volume) as well as through place promotion (Buck 1993; Kearns and Philo 1993; Kirshenblatt-Gimblett 1998; Hall and Page 2002; Hollinshead, this volume). For example, the 'discovery' of the Pacific by Europeans was the crucial point for the imaging of the Pacific. The early trading relationship with India and the Spice Islands of the Indonesian archipelago was an initial starting point into the creation of the image of the exotic. However, it was the accounts of French and English voyages of the seventeenth and eighteenth centuries which confirmed the discovery of 'paradise'. Contributing to this picture were two factors strongly influencing the Western mind in this period: the writings of Jean-Jacques Rousseau (1978) and the reassessment of Classicism, which had been stimulated by the unearthing of Herculaneum and Pompeii (Honour 1981). It was in islands of the Pacific that Rousseau's Romantic 'noble savage', elements of which had already been identified in the peoples of the Americas and south-east Asia, was to be discovered. Nevertheless, there were oppositional representations to the Romantic portrayal of the Pacific. Major Robert Ross, Lieutenant Governor of New South Wales, stated in 1788, 'I do not scruple to pronounce that in the whole world there is not a worse country . . . here nature is reversed' (in Hall 1992a). Similarly, the French explorer Baudin was aghast at the primitive nature of the Western coast of New Holland. 'In the midst of these numerous islands there is not anything else to delight the mind . . . the aspect is altogether the most whimsical and savage . . . truly frightful' (in Marshall 1968: 9). However, the Romantic image would come to predominate in Europe because of the means to which representations would be put. Official images of the Pacific emphasised the Romantic and the picturesque for two major reasons. First, such an image was in keeping with the dominant intellectual fashion of the times. Second, images could be put to utilitarian ends. Government utilised and encouraged such images in order to encourage settlement and imperialist mercantile and political expansion.

Concern over image and representation have become major concerns in some areas of tourism in recent years, but particularly with respect to the development of indigenous and so-called 'ethnic' tourism as well as heritage (Ashworth and Tunbridge 1996; du Cros, this volume; Fisher, this volume; Henderson, this volume; Marschall, this volume; Wels, this volume). For example, the term 'paradise' is often utilised in the promotion of postcolonial island states in a manner that reinforces Western ideas of a romantic other, in the same way that Eden has been applied to Africa (see Wels, this volume). In the case of Hawai'i, mercantile shipping connections between Hawai'i and the United States mainland served as the basis for both the annexation of the islands by the United States and the development of a tourism industry, to which commercial interests were applying the term 'Paradise' by the 1850s (Douglas and Douglas 1996). Yet, as Douglas and Douglas argued:

> The myth of Paradise is by now a thoroughly shop-worn cliché, which invests every kind of promotion . . . Virtually every travel brochure on the region contains similar images, no longer the exclusive preserve of Tahiti, which inspired them, or Hawai'i which mass produced them. By the 1970s, aided by jet travel, packaged vacations and the relentlessness of brochure and television advertising, the myth had been exported more widely than any other regional product and was being applied indiscriminately and often incongruously to every part of the Pacific.
>
> (Ibid.: 32–3)

Indeed, they went on to note that 'the myth had become so pervasive that its presence was evident even in the work of those who ought to be critical of it' (1996: 34) and illustrated this by noting that Farrell, in his introduction to *Hawaii: The Legend That Sells*, is lured to its use thus: 'Take a group of breathtakingly beautiful islands set in the blue Pacific as close to paradise as you wish . . .' (Farrell 1982: xiii).

Postcolonial issues of representation also include substantial examination of gender and the portrayal of women, as well as of race. Indeed, there is a substantial intersection of work between postcolonial studies and feminist and gender studies, including queer theory, with respect to the representation of other places and people (Gilman 1985; Kappeler 1986; Spivak 1986; Ware 1992; Blunt and Rose 1994; Ang 1995; McClintock 1995; Young 1995; Aitchison *et al.* 2002). One of the main reasons for such intersection is that concepts such as gender, class, ethnicity and race become a ground for 'internal colonialism' in which identities are constrained and oppressed and selectively represented. Because such constraints occur in the countries of both the colonised and the colonisers, such relationships are also often identified as providing common ground between the internally colonised of the First World and the externally and internally colonised of the postcolonial worlds (Childs and Williams 1997). Authors such as Spivak (1987) and Ashcroft *et al.* (1989) use this trope to describe how women in many societies have been relegated to the position of 'Other', marginalized and, in a metaphorical sense, 'colonized' (Spivak 1987). The postcolonial reading of gender issues and the representation of women has found substantial resonance in the study of tourism and gender issues, particularly with respect to the sexual exploitation of women and their representation in tourism advertising and promotion (Enloe 1989; Kinnaird and Hall 1994; Morgan and Pritchard 1998; Aitchison *et al.* 2002). (For an interesting tourism related subject that does not appear in the tourism literature see Ahmed (1998) and Wiegman (1999) for an account of tanning, skin colour and gender issues.)

The colonial and neocolonial dimensions of sex tourism has been explored in relation to the role of militarisation and the development of new international divisions of labour in which postcolonial relations are implicated (Enloe 1989, 1992; Hall 1992b, 1994a; Bishop and Robinson 1998). However, a significant theme within much of this research is the role of non-Eurocentric colonialism with respect to the role of the Japanese in south-east Asia prior to and during the Second World War and the subsequent development of sex tourism for the Japanese male market. Unfortunately, the potential implications of Japanese imperialism for the theoretical constructs of postcolonial studies are inadequately considered as are some of the implications of oppression within the patriarchal and racial structures of postcolonial countries themselves. Similarly, the phenomenon of female sex tourism is also not adequately theorised (Tucker 2003). Indeed, Hall (1992b) noted that in the south-east Asian context of sex tourism in the 1980s and 1990s many of the sex workers were from the internal periphery of those countries and often from ethnic minorities. In such a situation the institutionalised exploitation of women within patriarchal societies of south-east Asia has been extended and systematised by the unequal power relationship that exist not only between genders and members of ethnic groups but also between host and advanced capitalist societies (Ong 1985). A power relationship that extends to present-day gendered work practices (Oberhauser 2000) and the representation of women in tourism as subservient as well as sensual is perhaps best represented by Singapore Airlines ongoing promotion of 'Singapore Girl – a Great Way to Fly'.

The sexual imagery used in the marketing of certain postcolonial destinations such as the Caribbean or the Pacific (Opperman and McKinley 1997) tends to be a continuation of Western representations of a sensual, sexually available and subservient female oriental other since the seventeenth century (Hall 1998). Similarly, in the case of historical representations of Africans, Jordan (cited in McLintock 1995: 22) observed that by the nineteenth century, Africa was established as the quintessential zone of sexual aberration and anomaly in European lore as 'the very picture of perverse negation' that declared Africans to be 'proud, lazy, treacherous, thievish, hot and addicted to all kinds of lusts' (see also Wels, this volume). However, it is significant to note that McLintock regarded a focus on race or gender as singly defining categories for a sense of self as insufficient. Instead, she argued that gender is always racial and classed in the same way that race is always a gendered and classed category. Indeed, McClintock noted that it is precisely this interlinking and superimposition of categories that we should be analysing rather than assuming that gender, class or race are:

> distinct realms of experience, existing in splendid isolation from each other; they [cannot] be simply yoked together retrospectively like armature of Lego. Rather, they come into existence *in and through* relation to each other – if in contradictory and conflictual ways.
>
> (McClintock 1995: 5)

This is an observation that is only recently beginning to be incorporated into tourism studies (Aitchison *et al.* 2002; Tucker 2003).

Place, displacement and identity

The third major feature of postcolonial literatures identified by Ashcroft *et al.* (1989) is the concern with place and displacement and the 'special post-colonial crisis of identity'. Arguably the concern with identity in contemporary theorising is related not only to postcolonialism but also the postmodern condition. Displacement arises in postcolonial societies because of the processes of colonial settlement and migration, the transport of convicts, slaves or indentured labour, or by cultural denigration, whereby the indigenous culture is deliberately or even unconsciously oppressed by the colonial society (Bennett 1988), for example, in the case of deliberate policies of cultural assimilation of Aboriginal people in Australia until the early 1970s (Bennett 1988; Bhabha 1990; Ganguly 1992; Sharrad 1993). In locations of displacement, concerns over identity and authenticity occur as the identities of places and individuals come to be contested and renegotiated. Although language and text is critical to issues of identity, tourism also comes to play a major role in the construction of place and identities.

The role that tourism can play in transforming collective and individual values is inherent in ideas of commoditisation (Cohen 1977), which implies that what were once personal 'cultural displays' of living traditions or a 'cultural text' of lived authenticity become 'cultural products' that meet the needs of commercial tourism, as well as the construction of heritage. Such a situation may lead to the invention of traditions and heritage for external consumption that meet visitor conceptions of the other (Errington and Gewertz 1989; Buck 1993; Cronin and O'Connor 1993; Helu-Thaman 1993; Webb 1994; Ashworth and Tunbridge 1996; Chang *et al.* 1996; Picard and Wood 1997; Cronin *et al.* 2002; see also Fisher, this volume). However, it may be extremely difficult to distinguish between the creation of tradition for tourism versus its creation for other political and cultural ends of either the colonisers or the colonised (Hanson 1989; Keesing 1989; Trask 1991; Otto and Verloop 1996). Nevertheless, the very nature of the tourism industry may well create processes of acculturation and value change which are peculiar to tourism. For example, the imaging and marketing of destinations in tourism must commodify visitor and community notions of place and identity. As Papson commented:

> Tourism depends on preconceived definitions of place and people. These definitions are created by the marketing arm of government and of private enterprise in order to induce the tourist to visit a specific area . . . government and private enterprise not only define social reality but also recreate it to fit those definitions. This process is both interactive and dialectical. To the extent that this process takes place, the category of everyday life is annihilated.
>
> (1981: 225)

In the postcolonial setting, indigenous people may find themselves trapped, 'in a sort of tourized confinement in the suffocating straitjacket of enslaving external conceptions. They are caught in the objectifying slant of "Whites", "Westerners" and "Wanderers-from-afar" in an anonymous but continuing process of subjugation' (Hollinshead 1992: 44). For example, in the case of New Zealand, representations of Maori in tourist brochures have existed since the 1870s, while destinations such as Rotorua have long used aspects of Maori culture as a mechanism to attract overseas tourists (Carr 1999). With only a few notable exceptions (for example whale watching at Kaikoura), the tourism industry has long been *Pakeha* (European or outsider) dominated. As Barber commented:

> Pakeha New Zealanders have never been slow to exploit this indigenous culture in promotion and advertising – often in ways that drew Maori disapproval. There was a time when foreigners could have been excused for thinking, by the posters and videos they saw, that New Zealand existed solely of flax-skirted Maori jumping in and out of steaming pools.
>
> (1992: 19)

Nevertheless, it would be wrong to see postcolonial representations of identity as passively accepted by the colonised. Instead, cultural identity is 'an ongoing process, politically contested and historically unfinished' (Clifford 1988: 9). Tourism is clearly inseparable from such cultural politics, which can be defined as:

> the struggles over the official symbolic representations of reality that shall prevail in a given social order at a given time. One could argue that they are the most important kind of politics, for they seek to control the terms in which all other politics, and all other aspects of life in that society, will take place.
>
> (Ortner 1989: 200)

Therefore, tourism should not be observed in isolation as 'tourism inevitably enters a dynamic context, and in the process contention over definitions of what is traditional and authentic becomes charged with a variety of additional meanings, as the range of interested parties increases' (Wood 1993: 63–4).

'Cultural, ethnic, racial or national identities are commonly thought of as single, if not pure, markers of different locations within local and global society. . . . Hybridity calls attention to globalised persons and cultures and the condition of formerly colonized peoples' (Stoddard and Cornwall 1999: 332). Créolité, creolisation and métissage are often invoked as synonyms for hybridity and celebrated as non-hegemonic, open, creative processes that subvert the normative ideal of racial and cultural purity. Such concepts usually refer not just to the offspring of mixed-race or ethnic partners but also to the cultural mixing that occurs out of various forms of colonial encounter, including colonisation and globalisation but also, arguably, including tourism. For example, the Caribbean region is regarded as 'a key site of cultural hybridity whose centuries-old histories of dislocation and transmigration prefigure contemporary paradigms of globalisation and transnational, diasporic cultural identifications' (Stoddard and Cornwall 1999: 332) (see Duval, this volume, 2004b). Commentators, such as Bhabha (1994), argue that the hybrid subject offers the possibility of resistance to the totalising repression of colonialism (Hollinshead, this volume), while Papastergiadis (1997: 258) observes: 'The positive feature of hybridity is that it invariably acknowledges that identity is constructed through a negotiation of difference, and that the presence of fissures, gaps and contradictions is not necessarily a sign of failure.' Nevertheless, the appropriateness of the notion of hybrid identities for tourism is also being called into question by the application of the associated ideas of transnationalism and diaspora (Coles and Timothy 2004; Coles *et al.* 2004). As Hall (2003) commented, 'Given the context of globalisation, transculturalism and transnationalism we are all hybrids now'.

A recent significant departure for tourism studies is the examination of the interrelationships between tourism and migration (Hall and Williams 2002). Here, tourism may be seen related to the development of transnational communities in which the 'betweenness' of migrant belongings represents a significant strand of research (Fortier 2000; Duval 2004a, this volume). According to Coles *et al.* (2004) a transnational framework of analysis within tourism studies would allow for the recognition of interconnected social networks and the resulting movement between and among multiple localities. In other words, such interconnected transnational networks mean that movement, or temporary mobility, by transnational actors is perhaps another means by which tourism can be viewed. Such social networks and linkages may account for a significant amount of global tourism, especially when viewed in the context of migrant mobilities (Duval and Hall 2004). Nevertheless, the reality is that very little tourism literature has explored the tourist in the context of transnational behaviour (Duval 2003, this volume).

The dislocation and identity issues of postcolonialism are also to be found in recent interest in the intersections of tourism and diaspora (Shuval 2000; Coles and Timothy 2004; Coles *et al.* 2004). Clifford (1997), for example, noted that diasporic *routes* are as crucial in identity formation as the (geographical) *roots*. Notwithstanding inter- and intra-diasporic variations, Coles and Timothy (2004) observe that diasporas precipitate a number of different modes of travel and tourism inspired by the collision between their migrational histories (their 'routes'), their attachment to the 'home' country (their 'roots'), and their experiences of and in the host country (their 'routine'). Perhaps most predictably, diaspora tourists travel back to their original, ancestral homeland in search of their roots and family background (e.g. Bruner 1996; Stephenson 2002). More systematic, highly structured journeys of self-discovery that focused on the search for tangible artefacts of forebears have also been termed 'genealogical' (Nash 2002), 'family history' or 'ancestral' tourism (Fowler 2003). The second mode represents the first in reverse as residents of the original 'homeland' travel into the diaspora, while the third involves intra-diasporic travel to the far-flung destinations beyond 'home' occupied by diaspora(s). Fourth, spaces of transit in the scattering process along diasporic trajectories may assume sacred importance and motivate trips. For example, Ellis Island and the Statue of Liberty have become popular sites of pilgrimage for many European-Americans to pay homage to their forebears and their migrational achievements (Ashworth and Tunbridge 1996), although, as Ioannides and Cohen (2002) highlight, spaces of transit are not necessarily restricted to points of entry or departure. Finally, various diasporic home spaces and vacation spaces may also be significant for the decision to travel. One of the most interesting aspects therefore of recent writing on transnationalism, diaspora and mobility is that one set of movements leads to another. Displacement from one place to another and the subsequent implications that this has for identity may therefore mean that tourism becomes a perhaps unexpected beneficiary of postcolonial dislocation.

Postcoloniality and theory

The final key subject matter of postcolonialism is that of postcoloniality and theory. According to Ashcroft *et al.* (1989) the idea of postcolonial theory, literary or otherwise, has emerged because of the inability of European theories, themselves having emerged from particular cultural traditions which are hidden by false notions of 'the universal', to deal with the complexities and varied cultural provenance of postcolonial texts (Said 1984). A crucial thread to postcolonial theory therefore is that one of the aims of postcolonial analysis is not to assert a newly defined cultural power but to make visible the relative and partial nature of all 'truths'; and to

expose the ideological biases underwriting any ethical and epistemological system which would otherwise regard itself as definitive and axiomatic
(Nettlebeck 1992; Bahri 1995). Nevertheless, as Nettlebeck (1992) went
on to inquire: 'An important question to be asked about the post-colonial
project then becomes: to what extent does it disrupt or question constructions of political and cultural authority?' Indeed, Nettlebeck's question
raises the overall question of reliance on the 'colonial library' (Mudimbe
1994). As Finnström (1997) observed, colonialism and colonial hegemony
are not the only sources of power and cultural construction; the makers of
culture must not be limited to active colonisers, in the same way that local
populations must not be reduced to passive objects of cultural formation.
For Finnström (1997) this dichotomisation of active Westerners versus
passive non-Westerners, or 'givers' versus 'perceivers', is a long-lasting
misconception of Western thought that has unfortunately been a central
component of much postcolonial theory (Ashcroft et al. 1995; Bahri 1995),
although he went on to observe that the theoretical pitfalls of such analyses
are themselves highlighted by Ashcroft et al. (1995) when they note that
'The colonial space is therefore an agonistic space. Despite the "imitation"
and "mimicry" with which colonised peoples cope with the imperial presence, the relationship becomes one of constant, if implicit, contestation
and opposition' (Ashcroft et al. 1995: 9).

Therefore, in much postcolonial analysis 'the relationship of developers
and to-be-developed is constituted by the developers' knowledge and categories' (Hobart 1993: 2), in which local agents are presented as mere
objects to be changed (Hobart 1993: 14). As Finnström (1997) notes, theory
is then simplifying and homogenising actual postcolonial situations, and
therefore objectifying the postcolonial subjects, rather than the other way
round (see also Bahri 1995). Indeed, in a damning observation of the extent
to which Western Anglo-American academics dominate a field that paradoxically has the project of not asserting a space for colonial cultural
power, Majid argued that:

> postcolonial theory has been largely oblivious to non-Western articu-
> lations of self and identity, and has thus tended to interpellate the non-
> Western cultures it seeks to foreground and defend into a solidly
> Eurocentric frame of consciousness. Postcolonial theory thus operates
> with the paradoxical tension of relying on the secular, European vocab-
> ulary of its academic origins to translate non-secular, non-European
> experiences. Despite brilliant attempts to elucidate (or perhaps theo-
> rize away) this dilemma, the question of the non-Western Other's
> agency remains suspended and unresolved, while the material condi-
> tions that generate a culture of dubious virtues (such as 'hybridity' and
> 'identity politics') acquire more theoretical legitimacy. The question
> finally is: Will the subaltern be allowed to speak?
>
> (Majid 2001)

Similarly, Chabal observes:

> The present debate about our postcolonial identity is not one primarily concerned with the historical fact of the end of colonial rule (broadly from 1947 to 1964). There is indeed more talk today about the postcolonial than there was at the time of the end of empire. Nor is the postcolonial here meant to reflect the condition of African countries after independence. In the sense in which it is used in current cultural and ideological parlance, it refers to the implications of the postcolonial or postimperial condition of our own identity in the West today. *It is, therefore, more a concern about ourselves than about those who do live in actual postcolonial societies.*
>
> (1996: 37, editors' italics)

As Bahri (1995, 1997) and Chabal (1996) have highlighted, postcolonial studies too often ignore the actual societies of the postcolonies. In such a situation:

> Postcoloniality is divorced from the postcolony. Theory is then living a life of its own, without undergoing the critical contextualising of [ethnographic field research] . . . it is too essentialistic to write only in terms of binary oppositions of colonisers and colonised . . . no colonial state was working as a homogenous entity, they were all the result of a patchwork of conflicting and opposed social, political and economic interests.
>
> (Finnström 1997)

Therefore, the key binary categories in postcolonial theorisation, such as hegemony and resistance, must be complemented with aspects of localised strategies of adaptation, accommodation and collaboration (De Boeck 1996: 94).

Similarly problematic binary oppositions are prominent in tourism studies, such as tradition and change, and authentic difference and cultural homogenisation. Since the 'tourist gaze' (Urry 1990) is based very much on difference, discussion about the ways in which 'tradition' and 'authentic' cultures are impacted upon by tourism, and the ensuing need for measures of cultural preservation, are rife among a variety of tourism mediators and commentators. Postcolonial theory is useful in reminding us, however, that this aspect of tourism discourse which promotes the preservation of the 'traditional' for tourist experience is itself based on a colonial desire to fix the identity of the other in order that it remains (or perhaps in actuality becomes) distinct from tourist identity. Indeed, the global processes of tourism and modernisation do not necessarily erase notions of cultural authenticity because all global processes can only be understood specifically, and in terms of the premises of already existing cosmologies (Featherstone 1990). Accordingly, localisation is the inextricable concomitant of the globalising tendencies of postcolonialism and tourism. As Abu-Lughod (1991: 150) writes: 'the effects of extralocal and long-term processes are only manifested locally and specifically, produced in the actions of individuals living their particular lives, inscribed in their bodies and their words.'

This intersection between the global, the local and the individual then becomes a springboard not only for the interrogation of the subject of tourism but also for the development of a more theoretically informed tourism studies, of which an appreciation of postcolonialism will likely be a significant part. Nevertheless, the authors stress that to be relevant theorising must also be grounded in empirical research that seeks to contextualise theory development. Moreover, such an approach will inevitably make research more relevant to the subjects. Indeed, the emphasis on coloniser/colonised relations may serve to obscure the operation of internal political, social and racial oppression within the former colonies. While overall much postcolonial writing does not adequately acknowledge the rise of transnational capital, labour and culture and its implications for material realities of exploitation, as noted at the outset, the present book seeks to explore some of the contemporary ways in which the idea of the postcolonial can inform our understanding of tourism. It is to be hoped that the chapters contained in this book will help achieve such a purpose, and will also illustrate the potential of the tourism subject as a focus for postcolonial studies and new formations of transnationalism.

References

Abu-Lughod, L. (1991) 'Writing against culture', in R. G. Fox (ed.) *Recapturing Anthropology: Working in the Present*, Santa Fe, New Mexico: School of American Research Press, pp. 137–62.

Ahmed, S. (1998) 'Tanning the body: skin, colour and gender', *New Formations* 34: 27–42.

Aitchison, C., Macleod, N. E. and Shaw, S. J. (2002) *Leisure and Tourism Landscapes: Social and Cultural Geographies*, London: Routledge.

Ang, I. (1995) 'I'm a feminist but . . . "Other" women and postnational feminism', in B. Caine and R. Pringle (eds) *Transitions: New Australian Feminisms*, New York: St Martin's, pp. 57–73.

Ashcroft, B., Griffiths, G. and Tiffin, H. (eds) (1989) *The Empire Writes Back: Theory and Practice in Post-Colonial Literatures*, London: Routledge.

Ashcroft, B., Griffiths, G. and Tiffin, H. (eds) (1995) *The Post-colonial Studies Reader*, London: Routledge.

Ashworth, G. J. and Tunbridge, J. E. (1996) *Dissonant Heritage*, Chichester: Wiley.

Bahri, D. (1995) 'Once more with feeling: What is postcolonialism?' *ARIEL: A Review of International English Literature* 26, 1: 51–82.

Bahri, D. (1997) 'Marginally off-center: Postcolonialism in the teaching machine', *College English* 59, 3: 277–98.

Barber, D. (1992) 'Of tourism and tradition', *Pacific Islands Monthly*, August, p. 19.

Barnett, C. (1997) '"Sing along with the common people": politics, postcolonialism, and other figures', *Environment and Planning D: Society and Space* 15: 137–54.

Barnett, C. (1998) 'Impure and worldly geography', *Transactions of the Institute of British Geographers* 23: 239–51.

Batsleer, J., Davies, T., O'Rourke, R. and Weedon, C. (1985) *Rewriting English: Cultural Politics of Gender and Class*, London: Methuen.

Bennett, B. (ed.) (1998) *A Sense of Exile*, Perth: Centre for Studies in Australian Literature.

Best, L. (1968) 'A model of pure plantation economy', *Social and Economic Studies* 17, 3: 283–326.

Bhabha, H. (1984) 'Representation and the colonial text: a critical exploration of some forms of mimeticism', in F. Gloversmith (ed.) *The Theory of Reading*, Brighton: Harvester, pp. 93–122.

Bhabha, H. (1990) 'Dissemi/nation: time, narrative and the margins of the modern nation', in H. Bhabha (ed.) *Nation and Narration*, London: Routledge, pp. 291–322.

Bhabha, H. (1994) *The Location of Culture*, London: Routledge.

Bishop, R. and Robinson, L. S. (1998) *Night Market: Sexual Cultures and the Thai Economic Miracle*, New York: Routledge.

Blunt, A. and McEwan, C. (eds) (2002) *Postcolonial Geographies*, London: Continuum.

Blunt, A. and Rose, G. (eds) (1994) *Writing Women and Space: Colonial and Postcolonial Geographies*, New York: The Guildford Press.

Britton, S. G. (1982a) 'International tourism and multinational corporations in the Pacific: the case of Fiji', in M. J. Taylor and N. Thrift (eds) *The Geography of Multinationals*, Sydney: Croom Helm, pp. 252–74.

Britton, S. G. (1982b) 'The political economy of tourism in the Third World', *Annals of Tourism Research* 9, 3: 331–58.

Britton, S. G. (1983) *Tourism and Underdevelopment in Fiji*, Monograph No. 13, Canberra: ANU Development Studies Centre, Australian National University.

Britton, S. G. (1987) 'Tourism in Pacific island states, constraints and opportunities', in S. Britton and W. C. Clarke (eds) *Ambiguous Alternative: Tourism in Small Developing Countries*, Suva: University of the South Pacific, pp. 113–39.

Britton, S. G. (1991) 'Tourism, capital and place: towards a critical geography of tourism', *Environment and Planning D: Society and Space* 9, 4: 451–78.

Brockway, L. (1979) *Science and the Colonial Expansion: The Roles of the British Royal Botanical Gardens*, London: Academic Press.

Bruner, E. M. (1996) 'Tourism in Ghana: the representation of slavery and the return of the black diaspora', *American Anthropologist*, 98: 290–304.

Buck, E. (1993) *Paradise Remade: The Politics of Culture and History in Hawai'i*, Philadelphia, PA: Temple University Press.

Butcher, J. (2003) *The Moralisation of Tourism: Sun, Sand . . . and Saving the World?* London: Routledge.

Carr, A. (1999) 'Interpreting Maori cultural and environmental values: a means of managing tourists and recreationists with diverse cultural values', in Te Waipounamu, unpublished seminar paper, Centre for Tourism, University of Otago, Dunedin.

Chabal, P. (1996) 'The African crisis: context and interpretation', in R. Werbner and T. Ranger (eds) *Postcolonial Identities in Africa*, London: Zed Books, pp. 29–54.

20   *C. Michael Hall and Hazel Tucker*

Chang, T. C., Milne, S., Fallon, D. and Pohlmann, C. (1996) 'Urban heritage tourism: the global–local nexus', *Annals of Tourism Research* 23: 1–19.

Childs, P. and Williams, P. (1997) *An Introduction to Post-colonial Theory*, Englewood Cliffs, NJ: Prentice-Hall.

Clark, S. (ed.) (1999) *Travel Writing and Empire: Postcolonial Theory in Transit*. London: Zed Books.

Clayton, D. (2003) 'Critical imperial and colonial geographies', in K. Anderson, M. Domosh, S. Pile and N. Thrift (eds) *Handbook to Cultural Geography*, Thousand Oaks, CA: Sage, pp. 354–68.

Clifford, J. (1988) *The Predicament of Culture: Twentieth-Century Ethnography, Literature and Art*, Cambridge, MA: Harvard University Press.

Clifford, J. (1997) *Routes: Travel and Translation in the Late Twentieth Century*, Cambridge, MA: Harvard University Press.

Cohen, E. (1977) 'Toward a sociology of international tourism', *Social Research* 39, 1: 164–82.

Coles, T. E. and Timothy, D. J. (eds) (2004) *Tourism, Diasporas and Space*, London: Routledge.

Coles, T., Duval, D. and Hall, C. M. (2005) 'Tourism, mobility and global communities: New approaches to theorising tourism and tourist spaces', in W. Theobold (ed.) *Global Tourism: The Next Decade* (3rd edn), Oxford: Butterworth-Heinemann, pp. 463–81.

Connell, J. (1988) *Sovereignty & Survival: Island Microstates in the Third World*, Research Monograph No. 3, Sydney: Department of Geography, University of Sydney.

Craik, J. (1994) 'Peripheral pleasures: The peculiarities of post-colonial tourism', *Culture and Policy* 6, 1.

Crick, M. (1989) 'Representations of international tourism in the social sciences: sun, sex, sights, savings, and servility', *Annual Review of Anthropology* 18: 307–44.

Cronin, M. and O'Connor, B. (eds) (1993) *Tourism in Ireland: A Critical Analysis*, Cork: Cork University Press.

Cronin, M., Gibbons, L. and Kirby, P. (eds) (2002) *Reinventing Ireland: Culture, Society and the Global Economy*, London: Pluto Press.

Crosby, A. W. (1986) *Ecological Imperialism: The Biological Expansion of Europe, 900–1900*, Cambridge: Cambridge University Press.

De Boeck, F. (1996) 'Postcolonialism, power and identity: Local and global perspectives from Zaire', in R. Werbner and T. Ranger (eds) *Postcolonial Identities in Africa*, London: Zed Books, pp. 75–106.

de Reuck, J. and Webb, H. (1992), 'Postcolonial texts?: Introduction', *SPAN Journal of the South Pacific Association for Commonwealth Literature and Language Studies* 32 (wwwmcc.murdoch.edu.au/ReadingRoom/litserv/SPAN/32/intro.html).

Douglas, N. and Douglas, N. (1996) 'Tourism in the Pacific: historical factors', in C. M. Hall and S. Page (eds) *Tourism in the Pacific: Issues and Cases*, International Thomson Business Press, London, pp. 19–35.

Duval, D. T. (2003) 'When hosts become guests: return visits and diasporic identities in a Commonwealth Eastern Caribbean community', *Current Issues in Tourism*, 6, 4: 267–308.

Duval, D. T. (2004a) 'Conceptualising return visits: a transnational perspective', in T. Coles and D. Timothy (eds) *Tourism, Diasporas and Space: Travels to Promised Lands*, London: Routledge, pp. 50–61.

Duval, D. T. (ed.) (2004b) *Tourism in the Caribbean*, London: Routledge.
Duval, D. and Hall, C. M. (2004) 'Linking diasporas and tourism: transnational mobilities of Pacific Islanders resident in New Zealand', in T. Coles and D. Timothy (eds) *Tourism, Diasporas and Space: Travels to Promised Lands*, London: Routledge, pp. 78–94.
Edensor, T. (1998) *Tourists at the Taj: Performance and Meaning at a Symbolic Site*. London: Routledge.
Enloe, C. (1989) *Bananas, Beaches and Bases: Making Feminist Sense of International Politics*, Berkeley, CA: University of California Press.
Enloe, C. (1992) 'It takes two', in S. P. Sturdevant and B. Stolzfus (eds) *Let the Good Times Rool: Prostitution and the US Military in Asia*, New York: The New Press, pp. 22–7.
Errington, F. and Gewertz, D. (1989) 'Tourism and anthropology in a post-modern world', *Oceania* 60: 37–54.
Farrell, B. (1982) *Hawaii: The Legend that Sells*, Honolulu: University of Hawaii Press.
Featherstone, M. (ed.) (1990) *Global Culture: Nationalism, Globalization and Modernity*, London: Sage Publications.
Ferro, M. (1997) *Colonialism: A Global History*, London: Routledge.
Finnström, S. (1997) *Postcoloniality and the Postcolony: Theories of the Global and the Local*, Working Papers in Cultural Anthropology, No. 7, 1997, Department of Cultural Anthropology and Ethnology, Uppsala University.
Fortier, A.-M. (2000) *Migrant Belongings: Memory, Space, Identity*, Oxford: Berg.
Foster, S. (1990) *Across New Worlds: Nineteenth-Century Women Travellers and Their Writings*. London: Harvester Wheatsheaf.
Fowler, S. (2003) 'Ancestral tourism', *Insights* March: D31-D36.
Ganguly, K. (1992) 'Migrant identity: Personal memory and the construction of self-hood', *Cultural Studies* 6, 1: 27–50.
Gilman, S. L. (1985) *Difference and Pathology: Stereotypes of Sexuality, Race, and Madness*, Ithaca, NY: Cornell University Press.
Girvan, N. (1973) 'The development of dependency economics in the Caribbean and Latin America: review and comparison', *Social and Economic Studies* 22: 1–33.
Glage, L. (2000) *Being/s in Transit: Travelling, Migration, Dislocation*, Amsterdam: Rodopi.
Goldberg, D. T. and Quayson, A. (ed.) (2002) *Relocating Postcolonialism*, Blackwell, Oxford.
Hall, C. M. (1992a) *Wasteland to World Heritage: Preserving Australia's Wilderness*, Carlton: Melbourne University Press.
Hall, C. M. (1992b) 'Sex tourism in South-East Asia', in D. Harrison (ed.) *Tourism and the Less Developed Nations*, London: Belhaven Press, pp. 64–74.
Hall, C. M. (1994a) 'Nature and implications of sex tourism in South-East Asia', in V. H. Kinnaird and D. R. Hall (eds) *Tourism: A Gender Analysis*, Chichester: John Wiley, pp. 142–63.
Hall, C. M. (1994b) 'Ecotourism in Australia, New Zealand and the South Pacific: appropriate tourism or a new form of ecological imperialism?' in E. A. Cater and G. A. Bowman (eds) *Ecotourism: A Sustainable Option?*, Chichester/London: John Wiley/Royal Geographical Society, pp. 137–58.
Hall, C. M. (1998) 'Making the Pacific: globalization, modernity and myth', in G. Ringer (ed.) *Destinations: Cultural Landscapes of Tourism*, New York: Routledge, pp. 140–53.

Hall, C. M. (2003) 'Director's summary: Packaging the everyday culture of Canada/ L'empaquetage de la culture journalière du Canada', unpublished presentation, International Council for Canadian Studies Transculturalism conference, Montreal, May.

Hall, C. M. and Page, S. J. (2002) *The Geography of Tourism and Recreation* (2nd edn), London: Routledge.

Hall, C. M and Williams, A. (eds) (2002) *Tourism and Migration: New Relationships between Production and Consumption*, Dordrecht: Kluwer.

Hanson, A. (1989) 'The making of the Maori: culture invention and its logic', *American Anthropologist* 91, 4: 890–902.

Harrison, D. (ed.) (1992) *Tourism and the Less Developed Nations*, London: Belhaven Press.

Harrison, D. (ed.) (2001) *Tourism and the Less Developed Countries*, Wallingford: CAB International.

Helu-Thaman, K. (1993) 'Beyond hula, hotels, and handicrafts: a Pacific Islander's perspective on tourism development', *The Contemporary Pacific* 5, 1: 104–11.

Hitchcock, M., King, V. T. and Parnwell, M. J. G. (1993) 'Tourism in South-East Asia: introduction', in M. Hitchcock, V. T. King and M. J. G. Parnwell (eds) *Tourism in South-East Asia*, London: Routledge, pp. 1–31.

Hobart, M. (1993) 'Introduction: The growth of ignorance?', in M. Hobart (ed.) *An Anthropological Critique of Development: The Growth of Ignorance*, London: Routledge, pp. 1–30.

Hollinshead, K. (1992) '"White" gaze, "red" people – shadow visions: the disidentification of "Indians" in cultural tourism', *Leisure Studies* 11: 43–64.

Honour, H. (1981) *Romanticism*, Harmondsworth: Penguin Books.

Ioannides, D. and Cohen, M. W. (2002) 'Pilgrimages of nostalgia: patterns of Jewish travel in the United States', *Tourism Recreation Research* 27, 2: 17–25.

Johnston, R. J. (1991) *Geography and Geographers: Anglo-American Human Geography Since 1945* (4th edn), London: Edward Arnold.

Kappeler, S. (1986) *The Pornography of Representation*, Minneapolis: University of Minnesota Press.

Kearns, G. and Philo, C. (eds) (1993) *Selling Places: The City as Cultural Capital, Past and Present*, Oxford: Pergamon Press.

Keesing, R. (1989) 'Creating the past: custom and identity in the contemporary Pacific', *The Contemporary Pacific* 1: 19–42.

Kinnaird, V. and Hall, D. (eds) (1994) *Tourism: a Gender Analysis*, Chichester: John Wiley and Sons.

Kirshenblatt-Gimblett, B. (1998) *Destination Culture: Tourism, Museums and Heritage*. Berkeley: University of California Press.

Laurier, E. (1993) '"Tackintosh": Glasgow's supplementary gloss', in G. Kearns and C. Philo (eds) *Selling Places: The City as Cultural Capital, Past and Present*, Oxford: Pergamon Press, pp. 267–90.

Livingstone, D. N. (1992) *The Geographical Tradition: Episodes in the History of a Contested Enterprise*, London: Methuen.

Loomba, A. (1998) *Colonialism/Postcolonialism*, London: Routledge.

McClintock, A. (1995) *Imperial Leather: Race, Gender, and Sexuality in the Colonial Context*, New York: Routledge.

Mackenzie, J. M. (ed.) (1990) *Imperialism and the Natural World*, Manchester: Manchester University Press.

Majid, A. (2001) 'Provincial acts: The limits of postcolonial theory', *Post-ColonialismS/Political CorrectnesseS*, an international conference organized by The Postgraduate School of Critical Theory and Cultural Studies, University of Nottingham, and The British Council, Morocco, Casablanca, 12–14 April 2001 (www.postcolonialweb.org/poldiscourse/casablanca/majid1.html).

Marshall, A. J. (ed.) (1968) *The Great Extermination: A Guide to Anglo-Australian Cupidity, Wickedness and Waste*, London: Panther Books.

Matthews, H. G. (1978) *International Tourism a Political and Social Analysis*, Cambridge: Schenkman Publishing Company.

Meethan, K. (2001) *Tourism in Global Society: Place, Culture and Consumption*, London: Palgrave.

Morgan, N. and Pritchard, A. (1998) *Tourism Promotion and Power*, Chichester: John Wiley.

Mowforth, M. and Munt, I. (1998) *Tourism and Sustainability: New Tourism in the Third World*, Routledge, London.

Mudimbe, V. Y. (1994) *The Idea of Africa*, Bloomington, IL and Indianapolis, IN: Indiana University Press.

Nash, C. (2002) 'Genealogical identities', *Environment and Planning D: Society and Space* 20, 1: 27–52.

Nash, D. (1989) 'Tourism as a form of imperialism', in V. Smith (ed.) *Hosts and Guests: The Anthropology of Tourism* (2nd edn), Philadelphia, PA: University of Pennsylvania Press, pp. 37–52.

Nettlebeck, A. (1992) '"The two halves": Questions of post-colonial theory and practice in Christopher Koch's *The Year of Living Dangerously*', *SPAN Journal of the South Pacific Association for Commonwealth Literature and Language Studies* 32 (wwwmcc.murdoch.edu.au/ReadingRoom/litserv/SPAN/32/Nettlebeck.html).

Oberhauser, A. M. (2000) 'Feminism and economic geography: gendering work and working gender', in E. Sheppard and T. J. Barnes (eds) *A Companion to Economic Geography*, Oxford: Blackwell, 60–76.

Ong, A. (1985) 'Industrialisation and prostitution in southeast Asia', *Southeast Asia Chronicle* 96: 2–6.

Opperman, M. and McKinley, S. (1997) 'Sexual imagery in the marketing of Pacific tourism destinations', in M. Opperman (ed.) *Pacific Rim Tourism*, Wallingford: CAB International, pp. 117–27.

Ortner, S. B. (1989) 'Cultural politics: Religious activism and ideological transformation among 20th Century Sherpas', *Dialetical Anthropology* 14: 197–211.

Otto, T. and Verloop, R. (1996) 'The Asaro mudmen: local property, public culture?', *The Contemporary Pacific* 8, 2: 349–86.

Papastergiadis, N. (1997) 'Tracing hybridity in theory', in P. Werbner and T. Modood (eds) *Debating Cultural Hybridity: Multi-cultural Identities and the Politics of Anti-Racism*, London: Zed Books.

Papson, S. (1981) 'Spuriousness and tourism: politics of two Canadian provincial governments', *Annals of Tourism Research* 8, 2: 220–35.

Picard, M. and Wood, R. (1997) *Tourism, Ethnicity and the State in Asian and Pacific Societies*, Honolulu: University of Hawaii Press.

Pratt, M. L. (1992) *Imperial Eyes: Travel Writing and Transculturation*, London: Routledge.

Ranger, T. (1976) 'From humanism to the science of man: colonialism in Africa and the understanding of alien societies', *Transactions of the Royal Historical Society* 26: 115–41.

24 *C. Michael Hall and Hazel Tucker*

Rousseau, J. (1978) *The Social Contract and Discourses*, trans. G. D. H. Cole, rev. J. H. Brumfitt and J. C. Hall, Everyman's Library, London: Dent & Dutton.

Said, E. W. (1978) *Orientalism*, New York: Pantheon.

Said, E. W. (1984) *The World, the Text and the Critic*, London: Faber.

Selwyn, T. (1993) 'Peter Pan in South-East Asia: views from the brochures', in M. Hitchcock, V. T. King and M. J. G. Parnwell (eds) *Tourism in South-East Asia*, London and New York: Routledge, pp. 117–37.

Sharrad, P. (1993) 'Blackbirding: Diaspora narratives and the invasion of the body-snatchers', *SPAN Journal of the South Pacific Association for Commonwealth Literature and Language Studies* pp. 34–5(wwwmcc.murdoch.edu.au/ReadingRoom/litserv/SPAN/34/Sharrad.html).

Shuval, J. T. (2000) 'Diaspora migration: definitional ambiguities and a theoretical paradigm', *International Migration* 38, 5: 41–55.

Spivak, G. C. (1986) 'Imperialism and sexual difference', *Oxford Literary Review* 8, 1–2: 234–40.

Spivak, G. C. (1987) *In Other Worlds: Essays in Cultural Politics*, London: Methuen.

Spurr, D. (1993) *The Rhetoric of Empire: Colonial Discourse in Journalism, Travel Writing, and Imperial Administration*, Durham, NC: Duke University Press.

Stafford, R. A. (1984) 'Geological surveys, mineral discoveries, and British expansion, 1835–71', *Journal of Imperial and Commonwealth History* 12: 5–32.

Stephenson, M. (2002) 'Travelling to the ancestral homelands: the aspirations and experiences of a UK Caribbean community', *Current Issues in Tourism* 5, 5: 378–425.

Stoddard, E. and Cornwell, G. H. (1999) 'Cosmopolitan mongrel? Créolité, hybridity and "douglarisation" in Trinidad', *European Journal of Cultural Studies* 2, 3: 331–53.

Trask, H. (1991) 'Natives and anthropologists: the colonial struggle', *The Contemporary Pacific* 3: 159–67.

Tucker, H. (2003) *Living with Tourism: Negotiating Identities in a Turkish Village*, London: Routledge.

Urry, J. (1990) *The Tourist Gaze*, London: Sage.

Ware, V. (1992) *Beyond the Pale: White Women, Racism and History*, London: Verso.

Webb, T. (1994) 'Highly structured tourist art: form and meaning of the Polynesian Center', *The Contemporary Pacific* 6, 1: 59–85.

Wiegman, R. (1999) 'Whiteness studies and the paradox of particularity', *Boundary* 26, 2: 15–150.

Wood, R. E. (1993) 'Tourism, culture and the sociology of development, in M. Hitchcock, V. T. King and M. J. G. Parnwell (eds) *Tourism in South-East Asia*, London: Routledge, pp. 48–70.

Young, R. J. C. (1995) *Colonial Desire: Hybridity in Theory, Culture and Race*, London: Routledge.

Young, R. J. C. (2001) *Postcolonialism: An Historical Introduction*, Oxford: Blackwell.

2   Tourism and new sense

Worldmaking and the enunciative value of tourism

Keith Hollinshead

Introduction: the declarative value of tourism

In recent years work on the representative value of tourism has matured considerably, partly an account of the larger number of social scientists – from across the broad spectrum of the humanities – who have made sustained inspections of the subject. Indeed, the 1990s could be said to be the decade during which the declarative role or function of tourism came to be frequently traced and significantly understood. Put another way, it was the decade in which a broad spectrum of social scientists – generally working independently of one another rather than being part of any within-discipline/within-field 'push' – came to uncover and monitor the myriad ways in which tourism has been (and is) used to authoritatively announce or freshly affirm the felt 'true' character of places. Thus, it could be stated that during the 1990s considerable steps were taken by relatively detached or dissociated researchers into the partly conscious and partly subconscious ways in which tourism is used politically to articulate the so called 'real' nature of populations – that is, the preferred vision held by or about a particular local people.

Clearly, a number of individual theorists such as MacCannell (1976), Smith (1977), Richter (1980) and Cohen (1988) had dabbled productively on the inventive use by government bodies and corporate operators of tourism as a primary point of articulation of 'place', prior to the 1990s and such commentators have been well and deservedly applauded in within-field compendia such as Ritchie and Goeldner's (1987) massive tome, dedicated to the (International) Travel and Tourism Research Association – for which a second edition was produced in 1994. But the last decade of the twentieth century witnessed a steep incline in the number of social scientists who were able to mount substantial longitudinal inspections of the use of tourism to make, re-make and/or de-make specific peoples, places or pasts. It is helpful, here, to open this chapter (on the declarative value of tourism), to critique what these pioneer researchers on the articulative role and function of tourism have unearthed.

Theorising the declarative value of tourism: five thinkers

This section provides an introductory outline to the work on five of the leading theorists/thinkers (of the 1990s) on the declarative value of tourism. As such, it covers the ways in which we have come to understand how tourism is regularly used by various players/bodies/institutions to articulate preferred meanings of 'local' place and of – in its broadest sense – 'tribal' heritage. Each of the five researcher-commentators discussed have done much to advance our understanding of how the myths, the narratives and the interpretations which form the bedrock of travel-trade storylines and tourism-industry promotions are indeed routinely and heavily mediated or normalised in this fashion – much of it incrementally – before they are projected in and via the industry.

The first of the qualifying 1990s commentators is Elizabeth Buck, a fellow of the East-West Centre in Honolulu. Buck seeks to explore how the cultural and historical storylines of the tourism industry – acting in cahoots with other industries – does violence to the richness of Hawai'i's past experience. She actively traces – notably in music, chant and myth – how recent and contemporary representations of tourism and of the entertainment industry (predominantly) have constructed a new ideological apparatus within Hawaiian society, an apparatus that has assailed received manifestations of Hawaiian thought and experience. In this sense, Buck (borrowing from Sahlins 1983: 524) charts how the music and the narratives of tourism were manipulated as instruments of mytho-praxis to serve particular political interests in Hawai'i, frequently being part of a discursive order (after Foucault 1984) which was external to the existing/received logic of Hawaiian culture (Buck 1993: 56). Indeed, over time, Buck judges that tourism (an industry based on image) has functioned overridingly 'to construct, through multiple representations of paradise, an imaginary Hawai'i which [considerably] entices . . .' (Buck 1993: 179). Gradually, in Buck's view, 'almost everything in Hawai'i communicates through a system of codes that tourism, the public and private institutions that support tourism, have constructed over years of selling Hawai'i as paradise' (Buck 1993: 180).

Like Buck, the second of the lead commentators on the articulative value of tourism, Ian McKay, also investigated the politics of cultural selection. While Buck has probed the representation of 'paradise', McKay has been more concerned about how the tourism industry engages in the representation of 'innocence'. To him, decision-makers in tourism tended to work with a loose network of cultural producers in peddling distinct versions of an antimodern Nova Scotia; they were over-reliant on the authorial outlooks of particular folklorists and crafts-experts who had set themselves up as the natural/spontaneous voice of the so-called 'folk' of the province (McKay 1994: 99). Indeed, McKay even conceives of a folk formula at work in the mediation of place and the past. Under that folk formula a particular folk essence is 'found'. Then, in order, that essence is located within an esteemed 'Golden Age', that supposed Golden Age now represents a pastoral ideal, that pastoral ideal is called upon in the contemporary moment to inculcate singular visions of ethnic unity, and ultimately that essence-cum-Golden-Age-cum-pastoral ideal is thoroughly milked in the commodified search for profits (McKay 1994: 275).

In McKay's judgement, the provincial government of Nova Scotia gradually became increasingly enmeshed in the official production of such pastoral visions of innocence, to such a degree that a tourist state was created in governance. In various guises over time, the tourist state worked variously but concertedly with authority-craving folklorists (and then, later, with profit-hungry private corporations) to promote and sell 'folk products' abstractly, increasingly divorced from their real world contexts. Hence the tourist state comes to exploit 'the folk' in and via tourism past points of credibility. As Buck found for Hawai'i, tourists are encouraged to visit Nova Scotia to see/experience/collect things which have grown to become a sign of themselves.

Hal Rothman, the third theorist, is concerned (like Buck and McKay) at the distinctly axial role tourism can have across other industries and across society 'in special places' as it distends. To him, in the American West, the scale and scope of the influence of tourism to speak not only about places but for places is a severely misappreciated one. In his views, the very amorphousness of the phenomenon of tourism allows it to develop with little input, almost functioning autonomously. In this light:

> tourism is barely distinguishable from other forms of colonial economies. Typically founded by resident proto-entrepreneurs, the industry expands beyond institutional control, becomes institutionalised by large-scale forces of capital, and then grows to mirror not the values of the place but those of the traveling public.
>
> (Rothman 1998: 16)

In his inspection of tourism at the Grand Canyon, in Santa Fe, at Steamboat Springs and in Las Vegas, Rothman finds that tourism projects its own legible geography of and about places, and under the postmodern moment of our time, scripts the visitor at the centre of the picture. Through tourism, consumption thereby becomes an end in itself, and visitors to the American West have learnt not only how to consume tangible goods, but also the spirit and meaning of peoples, places and pasts. As such, the identity of various drawcard places in the American West ceases to follow what might have been the 'original' or 'traditional' iconography of those locations, but becomes a product of the international culture marketplace of tourism (Rothman 1998: 19). To Rothman, the scripting of drawcard destinations in tourism can fast become a faux chain, where the culture and the nature of revered spaces and places becomes a fount of psychic energy which not only affirms 'the nation' but affirms 'the visiting self' (Rothman 1998: 22). To Rothman, this is the evolving promise of tourism – an exercise in unrolling colonialism which impresses an American dreamscape or a similar homogenising and transformative spectacle upon and over such locales.

Publishing in the same year as Rothman, the fourth commentator that this section discusses, Barbara Kirshenblatt-Gimblett, also probes the degree to which tourism variously complicates or simplifies the experience of places, defining and redefining 'life' in destination areas. While Rothman talks of the industry's faux chain, Kirshenblatt-Gimblett suggests that the sorts of minimalised projections of and about 'destinations' which course through tourism in fact constitute a chain of performative effects. In her judgement, the institutionalised memory of tourism at leading heritage and travel destinations commonly tend to be theatrically mediated. The socially constructed projected 'madeness' and 'hereness' of places comprise a neat but totalised tourism realism – a ramified projection of place which may be deemed to be 'a performance epistemology' normalising ways of life for residents and 'a performance pedagogy' teaching visitors what is (or ought to be) important there (Kirshenblatt-Gimblett 1998: 194). Such are the reality effects of the mediations of representation in and through tourism: such is the aesthetic imperialism of the industry (Kirshenblatt-Gimblett 1998: 217). To Kirshenblatt-Gimblett, the potency of the industry to redefine places is much underestimated, particularly in terms of the very volume and the very velocity through which the dramatisations and the projections of tourism take place. As such, the tourism industry – in close alliance with the local/regional heritage industry – frequently constitutes a scarcely stoppable collaborative force which converts local places into extremely tightly scripted destinations. All too often the cultural inheritances of places not only lie in stewardship to the tourism industry but stand in problematic relationships to it. While Kirshenblatt-Gimblett acknowledges that there are many extraordinary experiments in experimental and virtual interpretation of places in tourism – such as at Plimoth Plantation (Kirshenblatt-Gimblett 1998: 189) – too many collaborative projections of place in tourism merely offer (as the articulated consciousness of places) banalised visions of 'difference'. Such scripted representations might frequently become hallucinatory for the residents of those places, yielding new or positive possibilities of life in and through actual or virtual performances of place. Moreover, all too regularly the interpretations of being which are indulged in are foreclosed: under such circumstances, the formal representation of culture has always belonged to the locally powerful and the so-called cultivated power-elite (Kirshenblatt-Gimblett 1998: 279).

The last investigator in this discussion of the declarative role and function of tourism is Keith Hollinshead, who is currently compiling a text on the value regimes and the related dominances and subjugations that come enwrapped within tourism industry and public culture industry representations. Drawing from the peer reviewed articles that Hollinshead has produced (e.g. 1998a, 1998b, 1999), over the past decade or more, Hollinshead's in-preparation text examines the fashions by which the tourism industry not only represents populations and revered cultural territories, but may be said to make, to de-make and to re-make those very locales. In this manner, the new text constitutes an iconology (rather than an iconography) for the industry, thereby reporting upon the role that tourism has played, and continues to play, in value transmission, in value development, and in value drift/value loss. Put another way, Hollinshead's emerging text constitutes an examination of the political character of the tourism imaginary, delving into the ways in which the aggrandising industry of tourism tends to inveigle itself in and among the myriad ways in which we see the world, we experience the world, we take meanings about the world, we know the world, and we thereby be (or exist) in the world. To these ends, the new *Tourism and Cultural Values* text entertains the sorts of value-issue debates which Hollinshead has written about during the 1990s, notably translating Foucault's work on the institutional apparatus of things to scenarios in tourism where particular agents-of-normalcy may be seen to be at work. It also draws significantly from the work of Horne (1992) on intelligent tourism: here, Horne had previously called for a healthier, more catalytic 'imaginary' across the tourism industry. His call for more 'intelligent' (i.e. inventively connective) forms of site revelation in tourism demands that the governments (in concert with the regional and national industry) of almost each and every nation ought to think much more creatively about what they do to make the held national 'cultural gene bank' viewable and understandable in stimulating, awe-inspiring, fashions.

Recap: recent research on the declarative value of tourism

The introduction to this chapter has been built around the authors identified above. It has drawn attention to the fact that in tourism studies during the 1990s, some impressive new longitudinal work has been conducted into the projective and population-defining/place-authorising role of tourism. For instance, from Buck has come evidence that the production of images (like 'paradise') can become an all-encompassing code by and through which the powerfully mediating tourism industry helps revalue things. From McKay has come the realisation that various levels of government work unsuspectingly hand-in-glove with many sorts of well-situated individuals, private institutions and corporate bodies in developing highly selective cultural frameworks determining what is thinkable, sayable or seeable 'locally'. From Rothman has come the recognition that tourism is 'a' if not 'the' principal vehicle of psychic or aspirational imagination. To

Rothman, tourism is the contemporary medium through which certain idio-syncratic populations and extraordinary places are manufactured in a layered and almost-patterned fashion. And, as destinations become increasingly popular and Aspenized (Rothman 1998: 354), original local residents frequently become internally colonised by the monied interests regulating tourism, increasing unable to lead participation in or control the operation of that industry.

Thereafter, from Kirshenblatt-Gimblett has come the judgment that tourism undoubtedly comprises the collaborative-consciousness industry for many places, today, and constitutes a mechanism of arbitrary and repetitive authentication which frequently freezes places within particular but limited visions of being and self-celebration. And from Hollinshead has come the synthesising view that tourism indeed serves today as the worldmaking medium of our time through which the poetics and aesthetics of our cultural and natural lives are politically contextualised (and de- and re-textualised) as particularly dominant visions of seeing and knowing are psychically naturalised and aspirationally commodified. Such is the generally objectifying, generally mainstreaming, and generally conservative agency of tourism – as the various researchers sometimes working within tourism studies but more commonly looking from outside, across tourism studies (from other disciplinary vantage points) are increasingly underscoring.

Tourism and postcolonial worlds

As Hall and Tucker have shown in the opening chapter to this book, there is a relative paucity of research on the interface between postcoloniality (or postcolonialism) and tourism (see Chapter 1). At this transitory or transitional moment for many of the globe's societies today, it is imperative that the conjugations between emergent/emerging/consolidating cultural groups and the world's so-called largest industry (Edgell and Smith 1994) is well measured and appropriately appraised. It is important to identify (if tourism is indeed an industry replete with Buckian powers of mediation, with McKayesque opportunities to culturally cleanse, and with Rothmanian influences to re-authorise places, etc.) the extent to which governing postcolonial states of various sorts have learnt to make use of the declarative reach of tourism to speak more frequently, more appositely, and more elegantly to their own kindled or re-kindled worlds. Hence, it is important to inspect the degree to which tourism is already being harnessed by individuals and institutions in charge of postcolonial states – or otherwise within postcolonial states – to address the substantive social, cultural and political problems of domination which have conceivably arisen through the recent experiences of colonialism, and through other attendant traumas of the colonial age.

Clearly, tourism may not yet be an industry ready for easy deployment by postcolonial states to connectively or refreshingly articulate things. Just as the novels and literature of the nineteenth and early twentieth centuries tended to fortify imperialism (Said 1993: 84), so the structure and condition of the late twentieth- and early twenty-first-century tourism industry may only axiomatically support the established 'imperial'/'colonialist-continuing' order. It may well be that 'tourism' and 'imperialism' are unavoidably mutually reinforcing entities, given the positional superiority of Western forms of consciousness at the helm of the industry (Meethan 2001: 42–9). Perhaps the tourism industry will always unavoidably be trapped by the need for immediate and therefore superficial communication about things, and will thereby always be a set of businesses which feeds off essentialisms (Thomas 1994), *eo ipso*. Perhaps tourism will always be the violence-rendering rhetorical instrument of imperialism, perpetually dealing in Eurocentric accounts (Hollinshead 1993a, 1993b) which tend to totalise the Western/North Atlantic view as the proper account for our received pasts and our lived presents. When Fanon (1965: 63) stated that 'Colonialism wants everything to come from [itself]', he was no doubt aware that globe-consuming Western travellers carried with them the dominant psychological features of the developed urban-industrial world and the accordant privilege to recognise/identify/position things in and of the world. Such, conceivably, is the routinely invasive hegemony of colonial/Western values through the everyday petty journeyings of tourists, and through the everyday petty actions of companies and corporations in the travel trade (Hollinshead 1999). Such is the contiguity between the Western knowledges of tourism and 'colonial' power (Urry 1990). Such is the universalising geography and the imperial memory of the exhibitorial force of tourism (Barringer and Flynn 1998). Such are the difficulties of life for the counter-narratives of the colonised and of the formerly colonised through the international, globalising architecture of the business of travel.

Now that tourism is increasingly being recognised for its power to articulate who a population is and to declare what ought to be celebrated about places (Horne 1992), it will increasingly be important to gauge what Gandhi (1998: 112) has called the discursive cartography of the field. In terms of postcolonial 'nations' and 'settings', it will be important to assess how the field of tourism is being made use of in the fresh or correct representation of particular societies. Hence, what value will tourism have as a field or a source of what cultural studies theorists and human communications specialists are calling 'utterance'. Indeed, it may now be expected that tourism – if Spivak's (1993: 56) words may be adapted – might indeed blossom 'into a garden where the marginal can speak'. In this regard, tourism could prove to be an important piece of armoury on the part of postcolonial states and populations in their efforts and freedoms to articulate the felt nationalisms and the cherished endearments which hold them together as 'people'. Thereby, tourism could/would/should prove to be a vital field through which revered or targeted 'strategic essentialisms' can be clarified and codified for internal consumption and otherwise announced and articulated for external digestion. This is the new role and function of tourism – to reveal the felt public culture of places. If one may borrow from Rushdie (1982: 116): 'A moment comes, which comes but rarely in history, when we step out from the old to the new; when an age ends; and when the soul of a nation long suppressed finds utterance . . . .' That is the new value for tourism: tourism in soul-work – tourism as the communicant of held 'being', particularly of suppressed and stifled 'being'.

In this and future decades, tourism will indeed play a large part in the articulation of the held 'imaginary' of places, and in communicating the recovered or expanding sensorium of populations. In postcolonial milieux, tourism will have a pivotal role to play in helping subjugated populations come to realise for themselves what had seemed impossible, and to attain what had appeared only 'imaginable'. In such settings tourism is a large international domain where embedded values and lost meanings can newly flower. Hence there will be an increasing need for many of those who work in tourism – in management positions as well as research teams – to be informed interpreters of 'being' and accomplished readers of text. Decidedly, those who will work in colonial encounter settings and post-colonial thinkers will need to be skilled at deciphering the political reach of competing or contesting textualities. If the narratives, the storylines and the interpretations of tourism are not only to purvey matters of meaning and being but to instigate matters of corrective or fresh 'becoming', those who work in tourism management and in tourism studies must not be blind to the ministry of textual agency. They must not be blind to the sorts of textual authority which runs through them as that very tourism textuality helps render or make the world.

In the everyday realm of postcolonial identity-making, where tourism will increasingly be seen as a most useful international vehicle of declar-ative articulation, the drawcards and displays of tourism will much more commonly be enlisted and engaged within available mechanisms of power. Those who work in tourism will need to allow for more complex under-standings about cultural space and cultural place. Clearly, as Gandhi (1998: 22) has realised, some postcolonial representations will still unavoidably stem from old/colonial narratives, while other postcolonial representations will stand as bright new counter-narratives. Yet nowhere will it be possible 'to return to or to rediscover an absolute pre-colonial cultural purity, nor [will it be possible] to create national or regional functions entirely independent of the historical implication in the European colonial enter-prise' (Ashcroft *et al.* 1989: 195–6). The emergent counter-textualities and cross-textualities of tourism will play on all sorts of negotiations, claims and aspirations. And some (in tourism) will even project resistance to the established civilising mission of contemporary tourism itself, being unwilling or unable to recognise the imaginative and declarative power of the subject.

Over the coming decades, those who work in tourism in postcolonial settings will need to come to terms with the new value in the international order of things of the non-West. For those trained and grounded in the West, the politics of knowing this non-Western 'Other' will be precarious and 'touchy', but for many a different breed of individual, it will also be vernal and exciting. The old textualities underpinning place-ness and nation-ness will be sometimes conjoined and sometimes replaced by new imaginary essences of place and a new diversity in the possibilities of collective/ national being. The emerging postcolonial fictionality of nationhood will produce – partly through the vocalisations of what we might call 'Declarative Tourism' – new local citizens and new political subjects. Tourism, as a discursive event, will help generate a new panoply in the rhetorics of futurity, and will help yield a new politics of style. Certain old coherencies about people, about places and about pasts will increasingly become suspect as new communities are imagined into being and called onto the international stage. If postcolonial heritage will be a history, and a presence, and a future in the making and baking, then those who work in tourism will have large culinary roles in the inventive manufacture of the produced world. Occasionally, significant new postcolonial or post-postcolonial worlds will unfold, partly articulated through the significations of international tourism. And those who work day by day in tourism will need to be much more vigilant to the new meaning systems and knowledge structures of the old periphery, or rather of the new pregnant South or the new irruptive non-West. Those who operate in tourism management and those who ferret in tourism research will no longer be able to afford to work through training and via qualifications tailored exclusively in terms of Western econometric prescription – as to some observers is potently the case today (Hall 1994; Meethan 2001). Unfolding forms of identity and affinity (King 1997) and new modes of difference (Hall 1997) will require much culturally subtle and politically nuanced responsiveness from those who work in tourism – which is conceivably the adolescent or the forthcoming 'Arch Business of Representation and Signification'. Are you ready for the new symbolic now in and through tourism? Are you prepared for the beckoning age of delineation and depiction per the medium of tourism? Let practitioners and researchers alike prepare themselves for the emergent textual geography of tourism, and for the emerging interpretive syllabary of tourism.

New sense in the postcolonial world: tourism, Bhabha and enunciation

Until the last couple of decades, postcolonial issues of being and belonging had been a somewhat featheredge subject within cultural studies, and the various situational and temporal aspirations of so-called postcolonial populations had been 'a notoriously difficult thing to write about [or to] be able to grasp' (Puranik 1994: 17). In recent years, however, the literary theorist

Bhabha has made rich contributions to critical thought on the postcolonial moment and its imperatives and impulses. Drawing pointedly from Said, Fanon and Foucault, in particular, his penetrating analyses of questions of identity and alterity under the postcolonial condition are reasoned at length in his well-received (if intense and microscopically nuanced) 1994 text *The Location of Culture*. This current chapter will now examine what tourism studies scholars – who conceivably study the quintessential business of 'difference-making' and 'other-making'! – can take from Bhabha's hairsplitting of cultural production and of emergent belonging. To that end, the chapter takes heavily from Hollinshead's two related interrogations of Bhabha's thinking for tourism studies, as are contained at length and in detail within Hollinshead (1998a, 1998b).

In most parts, *The Location of Culture* is a critical contemporary examination of the presuppositions, the discriminations and the under-examined partialities which essentialise, which naturalise and which discipline populations – especially 'other' populations – and which over time have concretised those people within canonical or imperious acts of signification. In *The Location of Culture*, Bhabha argues that postcolonial cultures and ethnicities, like all 'societies', are not as solidly 'distinct' or as pervasively 'particular' as is routinely observed. He maintains that many of the world's peoples – in rural-outcountry settings as well as in urban-industry locales – exist ambiguously in difficult or scarcely gauged third spaces, caught in awkward psychic circumstances in between well-known or vestigial forms of held difference. Thus, across each and every continent, and within each and every country, there are postcolonial populations and hybrid people who simply do not fit readily into the cultural/racial/ethnic classifications which not only government census officials provide for them but which are quietly built up in narratives about them in small-in-scale but large-in-consequence depictions, via tourism industry posters and via travel trade brochures. To Bhabha, the pain and torment caused by and through such often quiet and unthinking quotidian calumny can become incrementally and savagely crippling.

To Bhabha, esteemed social commentators such as Said had done much – especially in Said (1978) – to improve our understandings of cultural difference under the pressures of postcolonial existence, but too frequently significant findings had gravitated towards the Manichean division between a static/essential 'West' and an equally static/essential Other. In Bhabha's view, such polarised reasoning was far too reductionism and elemental, for the pursuit of bona fide cultural identification and bona fide inter-cultural engagement is much more knotty and entangled. Whenever Bhabha pries into the form and condition of supposedly alien but homogenous social realms, he generally finds there to be a veritable range of complex and intricate responses there. Notably at that moment when the representativity of colonial authorities begins break down, Bhabha tends to find evidence (in those fluid third spaces) of all sorts of re-emergent identity and newly emerging aspiration. To Bhabha, those multifarious forms of admixed culture and melding ethnicity are inevitably fused locations of cultural intensity: equally inevitably, they tend to be dialectical in the style and force of their articulation, and are thereby not only garbled in signification but temporally restless. To Bhabha, the identifications held by such third space populations will frequently appear to be perverse in their connectivities. Much deftness will be required to decently capture the seemingly inconstant and transitional identities issued by such halfway, restless or semi-heard, populations. At times, such in-between peoples will appear to mimic or imitate the cultural institutions of their erstwhile colonisers, and at times they may even appear to parody the supposed voices of their own precolonial 'Other'. It simply takes considerable interpretive skill and painstaking textual craft to fathom the (perhaps) schizophrenic and the (perhaps) ephemeral statements arising here and there from new, more relevant and longer lasting identifications. It takes a certain degree of embeddedness in local circumstance to determine which of the newly gelling or the reconsolidating codes-of-being and intensities-of-affiliation which flare up within these kinds of in-between communities really matter.

Hollinshead has worked with some perspicacity to decipher what Bhabha believes to be the emergent and ambivalent locations of culture in contemporary society, in the ongoing effort to translate Bhabha's ideas on cultural hybridity and ambiguity in the postcolonial moment to tourism and travel scenarios. The following ellipsian explanations of the Bhabhian term 'cultural hybridity' are taken from a number of such interpretations in Hollinshead (1998a: 132). Cultural hybridity is:

- that liminal space or interstitial passage between fixed identifications which entertains 'difference' without an assumed or imposed hierarchy – an expanded or ex-centric site of experience and empowerment;
- those productive Third Space articulations of cultural difference which reinscribe in-between spaces in international culture through cutting edge enunciations of translation and negotiation to thereby permit the people of those Third Spaces to elude the politics of polarity and emerge (i.e. to begin to re-envisage themselves) as the others of their selves;
- those sites of emergent cultural knowledge which resist unitary and ethnocentric notions of diversity, and which reveal culture to be uncertain, ambivalent and transparent, and open to the future;
- that space in between received rules of a priori cultural engagement where contesting and antagonistic forms of representation of culture stand on truths that are only ever partial, limited and unstable;
- that fantastic location of cultural difference where new expressive cultural identities continually open out performatively to realign the boundaries of class, of gender and of contingent upon the stubborn chunks of the incommensurable elements of past, totalised identity;

- those locations of social utterance which undergo historically transformative moments through the enunciation of 'inappropriate' symbolism to permit in-between peoples to contest these modernist understandings of being and identity which have hitherto tended to deprive them of their own subjectivities;
- those transnational and transitional encounters and negotiations over differential meaning and value in 'colonial' contexts where new ambivalent and indeterminate locations of culture are generated, but where that new celebration of identity consists largely of problematic forms of signification which resist discursive closure.

Hence, according to Hollinshead, Bhabha's mapping of the emergent and ambivalent locations of culture which he (Bhabha) finds to be so frequently housed in postcolonial settings is a recognition that such restless/in-between discourses of space and being iteratively interrogate (i.e. resist) the Western sense of synchronous tradition. These new-old and new-new flickering and/or ambiguous discourses repel the sorts of modernist and teleological consciousness of and about class, race and sexuality which Hollinshead (1993a, 1993b) and others (especially Crick 1989; Selwyn 1996; Sardar 1998; Meethan 2001) consider run quite rampantly through tourism management praxis and through tourism studies thinking. Clearly, then, the cultural-theoretical thought of Bhabha on 'differential peoples' and on 'interruptive cultures' has much to offer those who work in tourism in postcolonial scenarios. It is to be hoped that Bhabha's deep but refreshing metaphoricity about 'split locations' and 'fractured identities' can do much to help those who work in tourism studies demassify the supposed differences found in tourism/travel settings across the globe: conceivably, it can generate the gain of all sorts of new and more meaningful insights into the newness of identities and aspirations. The tourism studies 'academy' – with its scores of scholars regularly monitoring and calibrating the global business of 'difference' and the international machinery of 'othering', par excellence – must develop informed critical research agendas on the Bhabhian restlessnesses which in-between populations have to countenance. Over the immediate decades, the field of tourism studies must develop its specialists who can probe the syncretisms, the juxtapositions and the interminglings of cultural newness and of psychic selfhood. As Hollinshead (1998a: 125) has advocated, it now lies within the capacity of tourism, variously:

- to help emergent populations legitimise themselves, themselves;
- to 'produce' new politically resonant definitions of peoples, places and pasts through the tourism industry's everyday rhetoric and everyday communicative craft;
- to help partial or suppressed articulations of racial/ethnic/cultural identity survive;

- to empower certain populations in difficult predicaments (or sometimes in fortuitous circumstances) to fashion double or multiple identifications for themselves;
- to further contain or constrain other people within paradoxical scenarios where they are subject to vibrant counter-tensions of or about identity;
- to explore new possibilities of alterity within newly identified political-geographic spaces;
- to help previously suppressed community groups, or previously silenced ethnic populations, towards radically new representations of themselves which confidently contest mainstream or established delineations of them;
- to serve as a new channel for the performative projection of 'with-genre nostalgias' – something which is otherwise known as 'discursive remembering' (Bhabha 1990);
- to serve as a fertile field for the development of emergent cross-genre identifications of Selfhood (and of Otherness);
- to stand as a whole new medium through which subaltern peoples or emergent populations can experiment with the new lexicons of iconic identity as they creatively play at celebrating their felt 'new', or even their felt 'old', Selves.

Bhabha has warned us that peoples, places and pasts tend not only to be represented or misrepresented through substantial acts of articulation which are projected through grand events of signification. Rather more plainly, he suggests that they are miscast through what Foucault (see Morris and Patton 1979) had styled as *petits récits* – that is, through the apparently small everyday actions and the seemingly trifling events, but which ultimately cohere over time to solidify into a sedimented consciousness about that defined object (Bhabha 1994: 243).

Prospect: Bhabha and the worldmaking function of tourism

Bhabha's critiques on the hybridity and ambiguity of in-between populations – as contained in *The Locations of Culture* – helps one recognise the ubiquity and everyday commonality of the politics of articulation, and helps one readily comprehend how so many of the world's populations can and ought to be seen differently in their own emerging time of the now. The real merit in Bhabha's ongoing work on third space possibility is that a reading of *The Location of Culture* – should that not be the Locution of Culture! – emancipates understandings about cardinal matters of affiliative difference about life-force issues of macro-social inheritance, enriching our outlooks on what can conceivably be achieved in the reprojection of souls lost under the weight of external or misinformed representation. And a broad reading of and reflection about what Bhabha has to say about the new potential for fantasmatic projection promises (for restless populations) a whole new creative spectrum of options via tourism as 'a' or 'the' enabling speech act. And that broad reading of Bhabha should lead divergent thinkers to understand that it is not only postcolonial populations who stand as restless people – all of the world's populations, every ethnic group, every sub-cultural sort, every urban-industrial community is indeed inconstant, wakeful and agitated today: each and every self-declaring population needs its relinquishment from external representative oppression. There is simply no shortage of populations in the cosmopolitan North as well as in the so-called distant/peripheral South who could gain from what one might call 'enunciative uplift', whether it be through the re-significations of tourism or whatever other source of declarative agency is available.

Hence, this chapter has steadily emphasised the fact that tourism no longer ought to be seen as some singular or unconnected isolated-realm of the mere 'vacational' or of inconsequential 'leisure travel'. Tourism is a vital medium of being and becoming which not only talks about worlds, but decidedly makes (or, at least, helps make) worlds. Who was it that told us to 'beware . . . for when one writes the tourist brochure, one indeed defines the nation'?

Let us now retrace our steps a little in this chapter. In these important life-bearing and culture-conveying senses, we have recently learnt in tourism studies, from Buck, to beware of the constant meditative function of tourism – to beware of the pre-thinking which in fact pre-packages tourism storylines and even entire sites. Bhabha just makes us aware how absurd some of these pre-thought etic or ethnocentric notions can be as they enslave marginal people. Similarly, from McKay we have learnt how all kinds of under-suspected agents of the public and private sectors interfere with the history, the heritages and the hopes put in civic trust to them. Bhabha just makes us aware how small and seemingly insignificant those fragmentary mediations can be, yet how quickly they can cumulatively reduce suppressed folk to states of anomie. Similarly, from Rothman we have learnt that particular natural sites and particular cultural settings can have immense aspirational value for all sorts of populations who seek to bond with those sites by travelling to celebrate themselves at those very locales. Bhabha just makes us aware that societies do not aspire neatly and coherently in people-as-one blocks. Bhabha clarifies that the cultural heart-beat pounds differentially within communities, not just between them; the aspirational ache is inclined to be 'partial' rather than organic to whole, distinct, long-run societies – aspirations increasingly flicker along all kinds of new sub-cultural and sub-spiritual trajectories today. Similarly, from Kirshenblatt-Gimblett we have learnt how tourism is not only a highly performative fabricator of destinations but it is also a highly collaborative one where few players have the time or felt need to explore the foundational 'truth-quotient' of the natures, the heritages and the histories they system-ically peddle. Bhabha just makes us aware that that 'performative' and 'collaborative' need not always be negative in effect against the life-blood interests of bonded or bonding communities: it can almost just as easily be harnessed to issue positive or shock-value ameliorative significations of being and becoming – well, noting the weighted structure of the estab-lished industry, I did say 'almost'! And, similarly, from Hollinshead we have learnt that tourism does not just promote found environments and found cultures, it can decidedly de- and re-make the very world out there which we are able to touch, to see, to experience, to know and to spiritually revere. Bhabha just makes us aware that declarative agency, i.e. what Hollinshead later calls that worldmaking authority, is not restrictively corporate, nor is it exclusively 'Western', 'urban-industrial', or 'cosmo-politan': the imaginary worlds we wish to live in, to visit or to hail from are cosmologically open to the future. By way of a moot point, it is perhaps only the city of Jerusalem – the navel of so many different worlds – that is as yet satiated with competing kinds of affiliatory belonging!

Consequently, Bhabha's coverage of the creative force of new-sense articulations of being is of supreme relevance to those who operate within or who otherwise investigate within tourism. Granted, Bhabha's writing on ultra-refined topics such as 'postcolonial contramodernity' and 'anti-dialectics' will not be every social scientists' kettle-of-cultural-fish – particularly in the light of Bhabha's known long-paragraphed obscurities (Eagleton 1994). Granted, Bhabha's work rarely engages the material structures and the edifices by and through which creative enunciations have to be articulated and countervailing fantasmatics have to be exercised (Moore-Gilbert *et al.* 1997: 37). Granted, Bhabha lies relatively quiet on questions of class and gender, and is somewhat limited in the range of fields of 'othering' he embraces (p. 38). Granted, Bhabha's courage of the hybrid does not always move beyond the binary classifications he seeks to condemn (Young 1995). Granted, Bhabha tends to celebrate the power of discursive authority without frequently detailing the degree of self-consciousness that is necessary to orchestrate the political agency and the organisational activity behind those new visions of self effectively (Moore-Gilbert, *et al.* 1997: 38). And, granted that Bhabha may (for some observers in the non-West) appear nothing more than an accomplice of the Western intelligentsia, scarcely in touch with that which is culturally real and institutionally realisable in and from the (perhaps) marginal non-West (Ahmad 1997: 248). Yet, despite those important qualifiers, Bhabha does indeed destabilise our thinking about the regular, ordered and innate communities we supposedly live within.

Huntingdon (cited in Scruton 2002: vii) may have warned that Western universalism 'sees the whole world in terms of values that have their origin, meaning, and natural climate in what is in fact only a small (through admittedly noisy) part of it'. Certainly in tourism studies – a field still predicated on so many of those modernist universalisms (Meethan 2001), and a field over-concerned with the national slices of the tourist market and the vacation dollar – the most active and loudest voices continue to be 'Western'/'North Atlantic'. But Bhabha points directly ahead to the promise of all sorts of new energising cacophony up and down the post-colonial haunts and across and about our transcontinental diasporas. Are those who work in tourism ready for these new sorts of compounded utterance – for those sorts of discursive double-talk and restless speech? Bhabha himself did not recognise that tourism is and will increasingly be a key site for all sorts of these new cadences. To that extent he was rather deaf to, and unimaginative about, the new noises in and of tourism, *eo ipso*. But those in tourism studies who will take the time to read Bhabha and absorb his radar-signals about future-as-mixed and future-as-open fantasmatics will not be so dormant. They will fast learn, surely, that tourism is not just the operational business of travel journeying, ensuring that travellers arrive on time at the right destination. They will learn about the politic amplitude of tourism. They will learn that tourism will henceforward increasingly also be the identity business of textual negotiation, thereby helping one and all arrive at the right kind of new sense definitions for those sought or celebrated destinations.

Acknowledgements

The author acknowledges the careful help of Chunxiao Hou in the preparation of this chapter.

References

Ahmad, A. (1997) 'In theory: classes, nations, literatures', in B. Moore-Gilbert, G. Stanton and W. Maley (eds) *Postcolonial Criticism*, London: Longman. pp. 248–72.
Ashcroft, B., Griffiths, G. and Tiffin, H. (1989) *The Empire Writes Back: Theory and Practice in Postcolonial Literatures*, London: Routledge.
Barringer, T. and Flynn, T. (1998) 'Introduction', in T. Barringer and T. Flynn (eds) *Colonialism and the Object: Empire, Material Culture and the Museum*, London: Routledge, pp. 1–10.
Bhabha, H. (1990) *Nations and Narration*, London: Routledge.
Bhabha, H. (1994) *The Location of Culture*, London: Routledge.
Buck, E. (1993) *Paradise Remade: The Politics of Culture and History in Hawaii*, Philadelphia, PA: Temple University Press.
Cohen, E. (1988) 'Authenticity and commoditization in tourism', *Annals of Tourism Research* 15, 1: 29–46.
Crick, M. (1989) 'Representations of international tourism in the social sciences: sun, sex, sights, savings, and servility', *Annual Review of Anthropology* 18: 307–44.
Eagleton, T. (1994) 'Goodbye to enlightenment: Review of Homi Bhabha's "The Location of Culture"', *The Guardian*, 8 February.

Edgell, D. L., Sr. and Smith, G. (1994) 'International tourism policy and management', in J. R. B. Ritchie and C. R. Goeldner (eds) *Travel, Tourism, and Hospitality Research: A Handbook for Managers and Researchers*, New York: John Wiley, pp. 49–64.

Fanon, F. (1965) *A Dying Colonialism*, trans. by Haakon Chevaliar, New York: Grove Press.

Foucault, M. (1984) 'The order of discourse', in M. Shapiro (ed.) *Language and Politics*, New York: New York University Press, pp. 108–38.

Gandhi, L. (1998) *Postcolonial Theory: A Critical Introduction*, New York: Columbia University Press.

Hall, C. M. (1994) *Tourism and Politics: Policy, Power, and Place*, Chichester: John Wiley.

Hall, S. (1997) 'Old and new identities', in A. D. King (ed.) *Culture, Globalization and the World-system: Contemporary Conditions for the Representation of Identity*, Minneapolis, MN: University of Minnesota Press, pp. 41–68.

Hollinshead, K. (1993a) 'Encounters in tourism', in M. A. Khan, A. Olsen and T. Var (eds) *VNR's Encyclopaedia of Hospitality and Tourism*, New York: Van Nostrand Reinhold, pp. 636–51.

Hollinshead, K. (1993b) 'Ethnocentrism in tourism', in M. A. Khan, A. Olsen and T. Var (eds) *VNR's Encyclopaedia of Hospitality and Tourism*, New York: Van Nostrand Reinhold, pp. 652–62.

Hollinshead, K. (1998a) 'Tourism, hybridity and ambiguity: the relevance of Bhabha's "Third Space" cultures', *Journal of Leisure Research* 30, 1: 121–56.

Hollinshead, K. (1998b) 'Tourism and the restless peoples: A dialectical inspection of Bhabha's halfway populations', *Tourism, Culture and Communication* 1, 1: 49–77.

Hollinshead, K. (1999) 'Surveillance of the worlds of tourism: Foucault and the eye-of-power', *Tourism Management* 20: 7–23.

Hollinshead, K. (in prep.) *Tourism and Cultural Values: Ways-of-Viewing Peoples, Places, and Pasts*, Avon, England: Channel View Press.

Horne, D. (1992) *The Intelligent Tourist*, McMahon's Point, Australia: Margaret Gee Publishing.

King, A. D. (1997) 'Introduction: Spaces of culture, spaces of knowledge', in A. D. King, (ed.) *Culture, Globalization, and the World-System: Contemporary Conditions for the Representation of Identity*, Minneapolis: University of Minnesota Press, pp. 1–18.

Kirshenblatt-Gimblett, B. (1998) *Destination Culture: Tourism, Museums and Heritage*, Berkeley, CA: University of California Press.

MacCannell, D. (1976) *The Tourist: A New Theory of the Leisure Class*, New York: Schocken Books.

McKay, I. (1994) *Quest for the Folk*, Montreal: McGill University, Queens University Press.

Meethan, K. (2001) *Tourism in Global Society: Place, Culture, Consumption*, Basingstoke: Palgrave.

Moore-Gilbert, B., Stanton, G. and Maley, W. (eds) (1997) *Postcolonial Criticism*, Harlow: Addison-Wesley-Longman.

Morris, M. and Patton, P. (1979) *Michel Foucault: Power, Truth, Strategy*, Sydney: Feral Publications.

Puranik, A. (1994) 'Postcolonials from the edge: Review of Homi Bhabha's "The Location of Culture"', *Times Higher Education Supplement*, February, 11: 7.

Richter, L.K. (1980) 'The political uses of tourism: A Philippine case study', *Journal of Developing Areas* 14: 237–57.

Ritchie, J. R. B. and Goeldner, C. R. (eds) (1987) *Travel, Tourism, and Hospitality Research: a Handbook for Managers and Researchers*, New York: Wiley.

Rothman, H. (1998) *Devil's Bargains: Tourism in the Twentieth-Century American West*, Lawrence, KS: University of Kansas Press.

Rushdie, S. (1982) *Midnight's Children*, 2nd edn, London: Picador/Pan Books.

Sahlins, M. (1983) 'Other times, other customs: the anthropology of history', *American Anthropologist* 85.

Said, E. (1978) *Orientalism: Western Representations of the Orient*, New York: Pantheon.

Said, E. (1993) *Culture and Imperialism*, London. Chatto and Windus.

Sardar, Z. (1998) *Postmodernism and the Other: The New Imperialism of Western Culture*, London: Pluto.

Scruton, R. (2002) *The West and the Rest: Globalisation and the Terrorist Threat*, London: Continuum.

Selwyn, T. (ed.) (1996) *The Tourist Image: Myths and Mythmaking in Tourism*, Chichester: John Wiley.

Smith, V. (ed.) (1977) *Hosts and Guests: An Anthropology of Tourism*, Philadelphia, PA: University of Pennsylvania Press.

Spivak, G. (1993) *Outside in the Teaching Machine*, New York: Routledge.

Thomas, N. (1994) *Colonialism's Culture: Anthropology, Travel and Government*, Princeton, NJ: Princeton University Press.

Urry, J. (1990) *The Tourist Gaze: Leisure and Travel in Contemporary Societies*, London: Sage.

Young, R. J. C. (1995) *Colonial Desire: Hybridity in Theory, Culture and Race*, London. Routledge.

3 Saying the same old things

A contemporary travel discourse and the popular magazine text

Beverley Ann Simmons

Introduction

Inherent in any discursive study of contemporary travel representations are four distinct tensions surrounding tourism practices: the mass tourist and the autonomous traveller; reality and fantasy; sightseeing and embodied experience; and the present and the past. These nominated travel tensions, which researchers identify as particular theories of tourism, do not operate in a vacuum. Researchers have seldom brought them together in any coherent whole as a way to explain how these tensions are collectively worked together, or silenced, in specific combinations to produce a contemporary travel discourse. Foucault introduces the notion of conflicting discourses when he suggests that a society will favour a preferred version of truth over others, as a regime of truth, which in turn represses alternate versions of reality (1979). For the purposes of this chapter, discourse is defined as a social practice, which constitutes and conditions in its representation of power structures (see Foucault 1972; Wodak 1996). How do Western discourses stipulate specific regimes of truth about travel and about relations of power among travel constituents, including tourists, local inhabitants, place or mediators? How does a contemporary travel discourse depend on specific narratives about gender, race and class within a colonial discourse to stipulate such truths and relations?

An indication of how this discursive process operates can be found in cultural texts, such as in travelogues. These are usually published in the mass media, in travel sections of newspapers and popular magazines and in specialist travel magazines. Dann defines travelogues as evaluative and impressionistic post-trip accounts that are prepared for information, promotion and entertainment purposes (1992: 59). Travel journalists who write these accounts give selected and descriptive information about destinations and directions about tourist expectations. Travelogues consist of written and photographic content and usually cover between two and six pages of content. This chapter will examine how travel journalists as travel mediators (or those who frequently stand outside the tourist–local inhabitant relationship) construct power relations among travel constituents. The concept 'mediation' is defined in this chapter as an intervention between, and a shaping of, tourist relationships or interactions among diverse interest groups (Gunning 1998). Therefore, the aim of this chapter is to show how a contemporary travel discourse is arranged in written texts that promote international destinations for Western tourists; how a colonial discourse is essential to this arrangement; and how travel mediators use these discourses to construct a preferred version of contemporary, or postcolonial, travel practices in texts.

Several authors identify discourse analysis and travel mediation as gaps in travel research. Socio-linguistic and semiotic research, which has been used to examine written tourist texts such as brochures and guidebooks, has led to a growing interest in tourism language and discourse (see Urry 1990; Dann 1996; Duncan and Gregory 1999). Hollinshead proposes that tourism researchers must place more emphasis on understanding power relations within a discourse of travel and tourism (Hollinshead 1999, this volume; Jamal and Hollinshead 2001; see also Chambers 1997; Henderson 1992). Furthermore, there is a growing critique among post-structuralist feminists who encourage researchers to examine power and discourse, and especially in relation to Otherness, given that much of tourism research is preoccupied with the centrality of the Western tourist, and a masculine knowledge that determines travel practices (Aitchison 2001; Fullagar 2002).

Feminist scholars have undertaken considerable inquiry into women's travel writing during colonial expansion, demonstrating how women were marginalised by a colonial discourse (Mills 1991; Pratt 1992; Blunt 1994; Ker Conway 1998). For instance, Mills identifies a number of imperial and gendered discourses related to travel. First, the male, heroic adventure narrative of courage, strength, leadership and persistence perpetuated myths of empire. It situated adventure in remote settings away from domesticity or civilisation and outside the reach of women. Second, imperial discourses served to impose British order onto local inhabitants. To describe place is to master place, which Mills terms as a 'fantasy of dominance'. Third, a notion of Otherness created Europeans as a superior race, and colonised Others who were homogenised into a collective Other as inferior and who lacked moral scruples, decency or cleanliness (Mills 1991: 77–90).

Henderson, another literary analyst of travel writing, alleges that the landscape of travel is always being mediated through travel scripts (1992). Travel writers cannot escape from prior literature nor from literary traditions and genres that shape them as writers. Similarly, Stowe, in his analysis of guidebooks for nineteenth-century American travellers, shows how writers mediate by selectively constructing taste, safety, class and racial prejudices as they mediate power (1994). Further, Henderson (1992) argues that travel writing is designed to fire readers' imaginations, causing would-be tourists to follow others, whether real or fictional, alive or not (see also Stowe 1994). In this sense, temporal and spatial boundaries diminish as tourists not only enter a mediated relationship with place, but also a relationship with previous travellers.

Several researchers propose that travel writers draw from Western imaginations more so than from a knowledge of the Other (Bruner 1991; Henderson 1992; Bhattacharyya 1997; Fullagar 2002). For instance, mediation hides a number of Western myths, such as escapism. To escape is a mere illusion because, as Dubois and others allege, travel can only occur within the 'historical and cultural relations that arise out of colonial histories' (1995: 317; see also Butor 1992; Henderson 1992; Kowalewski 1992; Duncan and Gregory 1999). That is, Henderson (1992) says, we all carry our culture, experiences, expectations, skills, disciplines, memories, our self-identity and so on, with us.

A contemporary travel discourse in travelogues

A narrative analysis was undertaken to determine the shape of a contemporary travel discourse, its various elements and how they are fashioned to create a cohesive version of travel. This involved examining several travelogues that promoted destinations for international Australian tourists. These were published in four separate popular women's and travel magazines in Australia in 1998 and 1999 (see Simmons 2001, 2003). This analysis shows how a contemporary travel discourse operates within the four distinct domains of knowledge, emotions, senses and imagination. This discourse establishes a cohesive yet selective version of present-day Western travel practices and relations of power. It fixes meanings about travel: about relations among constituents and about cultural practices of travel. This discourse has its own set of social realities and discursive logic. A contemporary travel discourse is comprised of four discursive elements; privilege, desire and longing, sightseeing and fanciful play (see Table 3.1). Each of these elements is briefly defined as follows.

Privilege

An element of tourists' privilege presumes that tourists are already privileged within their Western culture. Narratives in these texts stipulate tourists' exclusive access to luxury and unspoiled nature or them being distanced from, or elevated above, local inhabitants and mass tourists. These are common strategies used to create tourist–Other relations: superior tourists and inferior local inhabitants and mass tourists. Another discursive pattern is to establish Western superiority over destinations through narratives that connect tourists with influences from colonialism or Western celebrities. These written texts construct an expectation among tourists that they are a travelling social elite and that places are already familiar and readily available to them.

Table 3.1 A contemporary travel discourse: the construction of travel constituents

An element of privilege identified in spatial narratives of the tourist's dominance and knowledge.

Tourists are socially elite in their own culture; dominant over place and others; and men are explorers and colonisers while women are dependent and romantic.

Tourists are not egalitarian and they do not seek relations with place and with others.

Place is assumed to be dangerous and needs to be mapped and made familiar.

Local inhabitants are diminished, disdained, denigrated or stereotypes.

Mediators are authorities on place and its features; they establish a specific version of tourism, as they map and interpret as familiar; and they invoke traces of western influence for tourists to follow.

An element of discovery identified in visual narratives of tourists' seeing, sightseeing, over other senses and experiences.

Tourists are sightseers, spectators; while other senses are diminished.

Tourists do not seek to engage all their senses in their travel; do not seek self-discovery; they are not experimental tourists.

Place is a spectacle; aesthetically beautiful and seductive; different, exotic, unusual, ornamental; and it is made famous by Westerners (historic or fantasy).

Local inhabitants are part of the scene and the spectacular; they are friendly, welcoming.

Mediators define sightseeing as *the* appropriate way to discover place, over other senses.

An element of desire identified in temporal narratives of tourists' desire and longing for the place of others.

Tourists long for an authentic past; the unchanged rural village and backwaters; where they want to encounter the past; to behave *as-if* locals or imagine themselves in past eras in timeless places; and they are hedonistic and desire romance, fantasy and pleasure in modernity.

Tourists do not desire to engage in a relationship with place and with others; to be socially interactive.

Place is seductive, romantic and it charms tourists; it is for tourists' indulgence and pleasure; yet it is also timeless, unchanging, where modernity is diminished.

Local inhabitants serve tourists' myths and fantasies.

An element of fanciful play identified in narrative of tourists' fantasy to not construct tourists as *real* travellers and in everyday realities as mass tourists.

Tourists need to escape into fantasies; from modernity, responsibility and social constraints; they are not constrained as mass tourists; and they travel *as-if* members of an elite class; colonisers in romanticised and imagined colonial or explorer relationships.

Tourists are not prepared to embrace realities, either their own or others, of travel practices.

Place is the tourists' playgrounds, sanctuaries from their everyday; found in islands, secluded backwaters and the . past

Local inhabitants are compliant and submissive to serve tourists and to sustain myths and fantasies.

Table 3.1 continued

Mediators' privilege – the past and tourists' relations with the past, and to erase the present.	Mediators create travel as play; teach tourists how to travel by diverting attention from everyday realities; from their dependency on travel's protective cocoon; construct tourists from colonial and male explorer discourses; and denigrate the mass tourist.

Desire

Tourists' desire for destinations is constructed in three dominant ways. First, desire is constructed when tourists are encouraged to seek those places that provide present-day pleasure, hedonism or escape from modernity. For example, in one travelogue, the South American city of Rio is constructed in stereotypes of hedonism and idleness, as a 'nirvana for the devotee of leisure' (Payne 1998). To further enhance Rio's hedonistic present, Rio's past is diminished when it is equated with decay and diminutive influences of Catholicism.

Second, desire is constructed as a longing or nostalgia for the past, as found in representations of agrarian and village life or regions untainted by modernity. For instance, in a travelogue text that promotes the Andalusian region of Spain as a destination, this region is said to exhibit a slower pace than modernity (Sheard 1998). In this example, the tourist is to seek out the domestic and past life as it is supposed to have existed in Spain's back regions. In this case, the present-time of Andalusian villages is erased, including the social conditions among local inhabitants, or between tourists and locals, to create a value for its past.

In fact, any place is likely to experience many past periods. However, it is a benign, romantic and peaceful version that is selected which hides manual labour that may have tamed harsh environments, domestic labour that transformed natural produce into home-made goods or warring factions among the villagers in other times. Influences from the impact of modernity or globalisation are silenced. Tourism is not presented as a solution to economic decline in this region. Demographic drifts that especially affect young people who seek work in the cities leave a predominantly older population in the peripheries. The remaining older population sanction the selected version of the past *as-if* they themselves make an 'authentic' past attainable for tourists.

The third category in an element of desire constructs a longing for colonial or Western influences. This is discussed shortly in the next section of this chapter.

Sightseeing

Of all the senses, sightseeing is presented as if this is the only way to know or discover place. Tourists are spectators and place is a spectacle. Tourists are out to see as much as they can in the time they have available. They collect symbols of status through seeing, which in turn, provides cultural currency at home. Place and local inhabitants are shaped in picturesque and exotic narratives that are designed to seduce tourists. Urry (1990) explains contemporary tourism as being preoccupied with the spectacle. He shows how the tourist gaze also establishes Western superiority within the tourist–Other relationship and alleges that self-discovery or self-transformation are lost in modern tourism (1990; see also Rojek and Urry 1997).

Fanciful play

The fourth identified discursive element of this discourse identified is fanciful play. Narratives of fanciful play draw the above three discursive elements of privilege, desire and sightseeing together, to transform the controlled mass tourist into an autonomous traveller. Illusions of freedom and escapism are preferred over realities that surround constriction and control of mass tourists. By pretending that tourists are not tourists but travellers, this discourse encourages tourists to ignore the reality of power relations within modern tourism, such as their dependence on the travel industry, or their relationships of power with locals as they might actually exist. The traveller narrative itself laments the loss of the ideal traveller to mass tourism. Furthermore, colonial discourses, with their inherent race, class and gender relations of power, such as colonial superiority and exploration, become important in this construction of a travel fantasy.

These four discursive elements all rely on the tourist being someone else, to occupy and see somewhere else, and to escape from some other place: a displacement that relies on power relations constructed as play – as a travel fantasy.

A colonial discourse as travel fantasy

A construction of present-day tourism as a fantasy depends on class, gender and race relations within a discourse of colonialism. Traces of colonialism and adventure can be shown to further liberate the tourist (see Kowalewski 1992; Pratt 1992; Duncan and Gregory 1999). This is what Bauman terms 'left over fantasies' that become integral to an assembly of travel illusions where he says 'the "real reality" has been already squeezed out . . . so that there inside, all is clear for the play called life: for life as play' (1994: 150). However, it is one version of colonialism that is used to accentuate the position of the dominant and ruling-class male coloniser and explorer. As Mills (1991) points out, this version romanticises selective colonial relationships as they might have been for some, though certainly not all, who lived under colonial rule. Colonial elitism further enhances the element of Western elite class privilege already bestowed on international tourists (see also Simmons 1999).

Two specific travelogue texts are examined here to show how a colonial discourse is constructed as fantasy. One text, about Barbados as an international tourist destination, demonstrates how narratives of colonialism and postcolonialism are worked together, while colonialism and exploration are combined in a text about Darjeeling. The Barbados text constructs place for tourists that is both modern (or postcolonial) and colonial (see extracts below). Specific postcolonial narratives explain how a colonial country has become postcolonial: cricket players and tourism legitimate a significant economic contribution by Barbados as a contemporary world player. Also, Barbados is synonymous with present-day male success and social prestige, when it is constructed as the 'finest' place in the world for cruising, sailing and diving – as an exclusive place for elite male leisure pursuits. In contrast, a colonial discourse is used to map its colonial past and influence. A prior colonial presence is presented as 'enjoyable' and beneficial to Barbados. In this text, tourists are expected to identify with a leisured lifestyle of the wealthy who visit or live in Barbados and with the aristocrats of former colonial life. Both are valued, as the following extract shows:

> Before gaining its independence in 1966, Barbados enjoyed 339 years of continuous British rule, which may explain its other great obsession: cricket. Tourism plays such an important role in the island's economy and there is no shortage of facilities . . . the Caribbean is considered to be among the world's finest cruising and sailing grounds . . .
>
> Well mixed with West Indian, the other most apparent racial element is African, from black slaves taken there in the early days to work the sugar plantations. For a glimpse of the colonial past, visit one of the old plantation houses. . . . You can imagine yourself part of that era of gracious living when the planters were rich, but at least you won't have to swelter as they did in unsuitably heavy clothes designed for the English climate. The house is furnished with gleaming antiques and there is a fascinating collection of tools and utensils formerly used in everyday life.
>
> (Cole 1999: 36–41)

Colonial class, gender and race relations are constructed *as-if* they are natural, while the oppression of colonialism is pushed to the background of this imaginative play. Tourists are aligned with the planter, who was most likely male, rich and an aristocrat, and not with the black African slaves who worked the plantations (also male). Tourists are not instructed to imagine themselves as the women in the planter's household, whether as the planter's family or as servants. In these two narratives of present-day and colonial privilege, fantasy elevates the tourist over an imaginary past, as well as over their contemporary social and financial status. To play *as-if* rich and famous, either in the past life as a British colonial or in contemporary elite leisure pursuits, are fantasies that are derived from elite male predilections. In this construction of travel, fantasy and place, tourists are in fact more superior than the colonial whom they are to emulate. This is evident when tourists are absolved from having to wear the heavy European clothing and from class, racial and gender oppression under colonial life, as it might actually have been lived.

Another example is shown in a text about Darjeeling as an international destination in which the travel author combines one discourse of the explorer–adventurer with another of the British colonial:

> Having grown up with images of the famous Sherpa Tenzing Norgay, it was not difficult to spot his grandson, Sonam Tshering, who was to be our host and guide. . . . There is, throughout India, a constant and persistent reminder of colonial times. . . . We loaded up the Maruti four-wheel-drive and headed off. . . . Soon we left the river plain . . . [and] began the four-hour, 2,000-metre climb towards Darjeeling. The road twisted its way relentlessly upward. . . . About half way up we stopped at a teahouse at Kurseong. . . . Like the tea, the view [of the Himalayas] was wonderful.
>
> [The Windamere Hotel] was a wonderful haven . . . [and] also has a wonderful history and is full of memorabilia. . . . At the Windamere expect log fires, hot water bottles, and deliberately maintained, old fashioned charm, not to mention the spectacular view across the mountain range. . . . Kalimpong itself looked like an English village based around a market square. I almost expected to see the Olde George and Dragon Inn.
>
> (Plummer 1998: 70–7).

In this example, the selected images about the tourist as a practising colonial and as a practising explorer effectively distance the tourist from local people, apart from the author's Sherpa guide. They provide a safe and exclusive space for tourists to practice this invented explorer–colonial narrative. The travel journalist, as a present-day explorer–colonial tourist, is advantaged because she can surround herself with familiarity in an unfamiliar place: with reminders that this place is well connected to her colonial origins. Narratives of an English colonial heritage lingering on India's landscape portray colonialism *as-if* it still exists in present-time for postcolonial tourists.

Tourists who follow this travel journalist can also participate in what Mills calls a 'fantasy of dominance' – to master place as male heroic adventurers (1991: 71). Their superiority over local guides who help Westerners to conquer mountains is a preferred narrative, rather than their dependence on local people at destinations. While Hansen (1999) identifies this superiority narrative in Western circulation from the mid-nineteenth century, more recent historical interpretations of this relationship challenge Western superiority, to construct a more egalitarian partnership between Western explorers and local guides. Despite this, the author reproduces a former version of those relations of power to establish her superiority as an explorer–colonizing tourist over her guide, the local Sherpa.

These contemporary travel texts are saturated with a colonial discourse. Fantasy is pivotal in a contemporary travel discourse to reproduce these images *as-if* they are the realities of modern mass travel.

How travel tensions are resolved

Travellers and not tourists

A new textual reality that offers the illusion of mass tourists' freedom is the ultimate travel fantasy: travel is *as-if* it frees the mass tourist. In this construction, travel is closed by a hegemonic discourse of *what is* – a superior tourist who has legitimate access to the space of others. This is based on a Western-centric discourse of *what was* – a colonial and travelling explorer who captured and controlled the space of others. Consequently, this travel discourse collapses a discursive tension between the tourist (as dependent and inauthentic) and the traveller (as being free, superior and authentic).

Reconfiguring the past into the present

Present-time tourism relations are erased in favour of past relations. That is, this discourse emulates the travel practices of the past: that a *traveller* is an explorer, and in so doing, adopts the narrative of the male privilege of colonial travel. This traveller is presumed to be free, autonomous and able to colonise place in a way that modernity precludes. Yet, this version of freedom and escapism itself may have been a fantasy, even in colonial times. To further uphold this fantasy of past colonial relations, the *traveller* adopts the social position of a colonial elite who does not interact with local people – declaring local people as inferior. When engagement with place and local people is denied in this discourse, due to class, racial and gendered superiority within colonial relations, the modern traveller becomes a sightseer precisely because there is no other way to encounter place and others. That is, the textual tourist-as-traveller, who is caught in a remote gaze of a colonial fantasy, is unable to make real contact with local people. This is why, Urry (1990) alleges, a discourse of present-day tourism lacks any capacity to transform tourists. Tourists need interactive contact with the Other to effect their self-discovery, or transformation (see Trinh 1994).

The contemporary travel discourse negates a desire among tourists to participate in the ordinary everyday lives of local people. Because this discourse constructs tourists to be out of touch with everyday realities in destinations, this means that the power of tourists is both expanded and diminished due to fantasy. Whereas tourists are socially superior surface-grazing tourists-as-travellers, seemingly, they do not impact on existing social or international relations when they are distanced from local inhabitants at destinations. This is quite contrary to imperial influences from colonial expansion and exploration.

Alternate travel discourses

In this preferred version of travel, several alternative discursive elements of travel are omitted or silenced by this dominant contemporary travel discourse. For instance, tourist–Other social relations are not constructed as egalitarian, and nor are tourists or local people at destinations expected to seek or establish interactive social contact. Tourists do not engage with place through a more fully sensory or embodied experience. Tourists do not embrace their own realities, the realities of Others or the realities of travel practices. Interestingly, a narrative of travel and self-discovery or as self-transformation through travel, which is one of the oldest tropes in the genre of travel writing, is missing from the texts examined.

Domesticity or exploration and conquest?

Ambiguities about women's travel are absorbed into this fantasy, resulting in women's invisibility in the travel text or their confinement within domesticity or romance. While the elements of male privilege and exploration from a colonial discourse are reproduced in a contemporary traveller discourse, it is only elite, white men who can explore, conquer or colonise, even if travel spaces are temporary, fanciful and visual possessions of place. Women tourists are presented with two courses for action within these discursive practices. First, they can adopt a travel narrative of male privilege, escapism from modernity, sightseeing and exploration, to become invisible by travelling *as-if* they are men. Second, they can adopt the feminised fantasy that transfers women from one domesticity into another. This trivialises women's travel when passive femininity is linked with domesticity in the domestic backwaters or romanticised past versions of place. These constructions link a discourse of femininity with travel, to privatise and culturally segregate the travel of women. In contrast, researchers of women's travel practices indicate that some women reject these elements identified as a contemporary travel discourse, to determine their travel in those discourses that are missing from the travelogue texts – in embodied, experiential or relation tourist discourses (see Veijola and Jokinen 1994; Wearing and Wearing 1996; Elsrud 1998; Simmons 2003).

Mediators and the reordering of power

Balancing fantasy against realities

Travel journalists, who reproduce a contemporary travel discourse in these texts, must balance illusions of freedom and escape against the political and economic demands of tourism organisations. When travel journalists select some elements of a contemporary travel discourse over alternative discourses or narratives of travel to construct the written travel text, they teach tourists how to be real travellers. Consequently, they camouflage political and economic demands when they favour fantasy illusions of escape and freedom. They reproduce tourists' privilege over local people, their desire for the past and past social relations, visual discovery as the way to know place, and fanciful play to hide travel's realities, rather than reproduce alternative discourses of relational, embodied experience or self-discovery. These alternative discourses may be more likely to expose travel's realities, such as a political and economic dependence within the structures of mass tourism.

Teaching tourists to travel

To balance illusions and demands, travel mediators teach tourists to surface graze. With one eye open, they encourage tourists to adopt the privileged status in the fantasy and to appreciate the spectacular beauty of place through the tourist gaze. With the other eye tightly closed, they encourage the tourist to shut out the realities of the everyday in its multiple forms. An illusion of tourists escaping can be kept alive when tourists are not anchored in the everyday realities of tourists and local people at tourist destinations. Integral to teaching tourists how to *travel*, travel mediators assign status and power to the tourist in four distinct ways when fantasy is used to prevail over reality. First, tourists are elevated above their own contemporary socio-economic status and the status of others within their home culture and above the local inhabitants and other tourists at tourist destinations. Therefore, status provides licence for them to raid and invade the everyday worlds of others *as-if* destinations belong to them. Second, fantasy seduces tourists to become *as-if* colonisers who are in fact more privileged than real colonisers in selectively romanticised colonial relationships. Third, fantasy constructs tourists *as-if* they are explorers on roads *as-if* they are less travelled. Freedom and exclusivity on these roads are themselves an illusion.

Finally, the promotion of fantasy also presumes that there are no distinctions between the way that men and women travel, as members of a privileged elite class. While travel is constructed to presume a male traveller, women's travel is either invisible or trivialised in this discourse. In summary, mediators teach present-day tourists to replicate past versions of travel in their travel practices, as adventurers, explorers and colonisers, as men once did – from a pre-determined travel fantasy script. The fantasy and desires of contemporary travel constitute tourists at play: the mediators are their instructors, while the everyday worlds of the Other are their playgrounds.

Conclusion

Those travel tensions nominated at the beginning of this chapter – namely, autonomous travellers and mass tourists, the present and the past, fantasy and reality, and sightseeing and embodiment – are diminished when a preferred regime of truth is established in a contemporary travel discourse. That is, tourists *become* travellers; the past *erases* the present from among local inhabitants at destinations, whereas the present *foregrounds* tourists' hedonism; fantasy *becomes* the reality of modern travel; and tourists, who are caught in the social relations of a colonial time-warp, must *become* sightseers.

This textual analysis of the written travel text in Western popular magazines shows that the remains of imperialism not only linger in Western imaginations about cultural, racial and gender superiority but are in fact central to a contemporary travel discourse that is reproduced in the written travelogue text. The postcolonial Western-centric imaginations about cultural elitism, hedonistic desire, freedom and escape, as well as colonial and exploration discourses with their inherent race, class and gender relations of power, are essential to construct a regime of truth about travel practices and social relations in a contemporary travel discourse.

Travelogue writers reproduce the same old things about the autonomous traveller, though very little about the social context of present-day, or postcolonial, travel relations and practices. Preferred romanticised and sanitised versions of colonialism, exploration and destinations allow Western tourists to pretend that even under the conditions of mass tourism, they can still attain their Western-centric imaginations of elitism, freedom, escape and conquest. Thereafter, according to this discourse, they sustain racial, class, gendered and cultural elitism. This backward-looking playful representation of travel, which is identified in the travel texts in popular magazines, constructs a version of travel truth which silences tourists' engagement with place and with others. That is, travel is not constructed as a fertile social space. The power structures of a contemporary travel discourse inhibit any potential for present-day tourists and local inhabitants at destinations to imaginatively forge interactive and profound social relationships. How present-day international tourists, whether Westerners or not, negotiate, dismantle, resist or sustain these multiple, yet seemingly cohesive, elements that comprise a contemporary travel discourse in their travel practices is a fertile ground for further tourism research.

References

Aitchison, C. (2001) 'Theorizing Other discourses of tourism, gender and culture: can the subaltern speak (in tourism)?', *Tourist Studies* 1, 2: 133–47.

Bauman, Z. (1994) 'Desert spectacular', in K. Tester (ed.) *The Flaneur,* London and New York: Routledge, pp. 138–57.

Bhattacharyya, D. P. (1997) 'Mediating India: an analysis of a guidebook', *Annals of Tourism Research* 24, 2: 371–89.

Blunt, A. (1994) *Travel, Gender, and Imperialism: Mary Kingsley and West Africa,* New York and London: The Guildford Press.

Bruner, E. M. (1991) 'Transformation of self in tourism', *Annals of Tourism Research* 18: 238–50.

Butor, M. (1992) 'Travel and writing', in M. Kowalewski (ed.) *Temperamental Journeys: Essays on the Modern Literature of Travel,* Athens, GA: University of Georgia Press, pp. 55–68.

Chambers, E. (ed.) (1997) *Tourism and Culture: An Applied Perspective,* Albany, NY: State University of New York Press.

Cole, L. (1999) 'A little taste of Britain', *Travel Away,* January, pp. 36–41.

Dann, G. (1992) 'Travelogs and the management of unfamiliarity', *Journal of Travel Research* 30, 4: 59–63.

Dann, G. (1996) *The Language of Tourism. A Sociolinguistic Perspective,* Wallingford: CAB International.

Dubois, L. (1995) '"Man's darkest hours": Maleness, travel and anthropology', in R. Behar and D. A. Gordon (eds) *Women Writing Culture,* Berkeley and Los Angeles, CA, and London: University of California Press, pp. 306–21.

Duncan, J. and Gregory, D. (eds) (1999) *Writes of Passage: Reading Travel Writing,* London and New York: Routledge.

Elsrud, T. (1998) 'Time creation in travelling: the taking and making of time among women backpackers', *Time and Society* 7, 2: 309–34.

Foucault, M. (1972) *The Archaeology of Knowledge,* trans. S. Smith, London: Tavistock.

Foucault, M. (1979) 'Truth and power: an interview with Alessandro Fontano and Pasquale Pasquino', in M. Morris and P. Patton (eds) *Michel Foucault: Power/Truth/Strategy,* Sydney: Feral Publications, pp. 48–58.

Fullagar, S. (2002) 'Narratives of travel: desire and the movement of feminine subjectivity', *Leisure Studies* 21, 1: 57–74.

Gunning, T. (1998) 'The whole world within reach: travel images without borders', in C. T. Williams (ed.) *Travel Culture: Essays on What Makes us go,* Westport CN: Praeger, pp. 25–38.

Hansen, P. (1999) 'Partners: guides and Sherpas in the Alps and Himalayas, 1850s–1950s', in J. Elsner and J. Rubies (eds) *Voyages and Visions Towards a Cultural History of Travel,* London: Reaktion Books, pp. 220–31.

Henderson, H. (1992) 'The travel writer and the text: "My giant goes with me wherever I go"', in M. Kowalewski (ed.) *Temperamental Journeys: Essays on the Modern Literature of Travel,* Athens, GA: University of Georgia Press, pp. 230–48.

Hollinshead, K. (1999) 'Surveillance of the worlds of tourism: Foucault and the eye-of-power', *Tourism Management* 20: 7–23.

Jamal, T. and Hollinshead, K. (2001) 'Tourism and the forbidden zone: the underserved power of qualitative inquiry', *Tourism Management* 22: 63–82.

Ker Conway, J. (1998) *When Memory Speaks: Reflections on Autobiography,* New York: Alfred A. Knopf.

Kowalewski, M. (ed.) (1992) *Temperamental Journeys: Essays on the Modern Literature of Travel,* Athens: University of Georgia Press.

Mills, S. (1991) *Discourses of Difference: An Analysis of Women's Travel Writing and Colonialism,* London: Routledge.

Payne, J. (1998) 'Rio rhythm', *Australian Vogue Entertaining and Travel,* September, pp. 116–21.

Plummer, A. (1998) 'High Tea', *Travel Away,* October–November, pp. 70–7.

Pratt, M. L. (1992) *Imperial Eyes: Travel Writing and Transculturation,* London: Routledge.

Rojek, C. and Urry, J. (eds) (1997) *Touring Cultures: Transformations of Travel and Theory,* London and New York: Routledge.

Sheard, J. (1998) 'The villages of Andalusia', *Australian Vogue Entertaining and Travel,* October, 128–30.

Simmons, B. A. (1999) 'The travelog in popular women's magazines', in M. Foley, M. Frew and G. McPherson (eds) *Leisure, Tourism and Environment: Participation Perceptions and Preferences,* Eastbourne: Leisure Studies Association.

Simmons, B. A. (2001) 'Tracing travel talk in the popular magazine text', *Annals of Leisure Research* 4: 77–94.

Simmons, B. A. (2003) 'Travel talk: When knowledge and practice collide', unpublished Ph.D. thesis, Newcastle, NSW: The University of Newcastle.

Stowe, W. W. (1994) *Going Abroad: European Travel in Nineteenth-century American Culture,* Princeton, NJ: Princeton University Press.

Trinh, T. Minh-ha (1994) 'Other than myself/my other self', in G. Robertson, L. Tickner, J. Bird, B. Curtis and T. Putnam (eds) *Travellers' Tales: Narratives of Home and Displacement,* London: Routledge, pp. 9–26.

Urry, J. (1990) *The Tourist Gaze: Leisure and Travel in Contemporary Societies,* London, Newbury Park, and New Delhi: Sage Publications.

Veijola, S. and Jokinen, E. (1994) 'The body in tourism', *Theory, Culture and Society,* 11: 125–51.

Wearing, B. M. and Wearing, S. L. (1996) 'Refocusing the tourist experience: The flaneur and the choraster', *Leisure Studies,* 15, 4, pp. 229–44.

Wodak, R. (1996) *Disorders of Discourse,* London: Longman.

4 Cultural tourism in postcolonial environments

Negotiating histories, ethnicities and authenticities in St Vincent, Eastern Caribbean

David Timothy Duval

> I have put quotation marks around the word native because, as I discuss below, they rarely are.
>
> (Forster 1964: 224)

Introduction

The search for 'the authentic' (as opposed to 'authentic'), some would argue, motivates many modern tourists from Western countries to visit postcolonial environments where the 'lost' Other (Shepherd 2002; Vivanco 2002) can be found. It is this historical connection that serves as the foundation for the surge of cultural tourism in many of these environments and localities. The relationship between authenticity and the tourist experience has been extensively discussed in the literature (see, for example, MacCannell 1976; van den Berghe 1994; Cohen 1995; Hughes 1995; Edensor 1998, 2001). It has been largely focused on generating a primarily etic assessment (yet occasionally affording an emic, or insider's, perspective) of the validity of those cultural expressions and patterns presented for tourist consumption. The assertion often made is that tourists do not experience authentic culture and that they are, consequently, witness to 'staged authenticity' (MacCannell 1976) or a 'pseudo-event' (Boorstin 1964). Through Urry's (1990) 'gaze', tourists seek and consume the sublime yet unfamiliar. Under the umbrella term of cultural tourism, consideration is frequently given to how cultures and social organisation are affected by the presence of tourists and their gaze. At issue in the study of cultural tourism is how social and cultural identities are managed and negotiated in the touristic environment in the face of the tourist gaze, and subsequently how such processes relate specifically to various types of cultural tourists (Titley 2000; Tucker 2001; McKercher 2002). Of course, it is often the cultural tourism product that is the subject of debate over its 'authenticity', particularly in the form of performances, ethnic identities (or ethnicities) and tradition. Of more interest, I would argue, are the processes and underlying historical context(s) upon which the performance or broader cultural identity is ultimately based.

The purpose of this chapter is to offer an alternative angle from which one might view staged authenticity, or more specifically, constructed realities in the context of tourism and touristic activities. It is argued that such an angle would incorporate and situate historically the object, tradition, ethnicity or 'performance' (Edensor 2001) being presented for consumption. I suggest that an examination of specific and overt elements associated with cultural tourism (e.g. performances and souvenirs) requires an acknowledgement of the specific histories of culture and tradition being presented. Whether or not these histories are 'authentic' depends on the perspective taken, who is actually doing the gazing and even who is involved in orchestrating and packaging the gaze itself. My aim, therefore, is to explore the intersection between histories as culturally constructed narratives, ethnic identities and ethnicity, and the development and resulting product of cultural tourism. I argue that, in order to incorporate such historical contexts, an approach is needed that utilises an ethnohistorical/historiographic approach, which can ultimately allow for the contextualisation of a particular social or cultural element being offered to the tourist for regulated consumption. In other words, one needs to be conscious of the larger historical processes that have shaped the very performance, culture or attribute under scrutiny. Tied in with this is how 'tradition' is conceptualised in a touristic context, largely because what is often presented as authentic (even if intentionally staged) is inherently steeped and rationalised on the basis of tradition(s) that are, themselves, directly linked to specific (and often multiple) social and cultural histories.

By way of an example, recent tourism development initiatives involving the 'Carib', an indigenous group living in the northern coastal regions of St Vincent in the Eastern Caribbean, are explored. These initiatives, it is shown, were initially fostered under the guidance of action research undertaken by Franklin (1993) and Franklin and Morley (1992) that was designed to instigate community-based 'nature tourism' that ultimately incorporated cultural elements. I argue that such efforts, while admirable from the perspective of community-centred drives for economic empowerment involving tourism, have helped shape for consumption not *the* 'Carib' culture but rather an historically *filtered* 'Carib' culture (see also Lacy and Douglass 2002).

As cultural tourism often incorporates the commodification (and its often negative connotation) of ethnic and social identities (if not as the direct object for consumption then perhaps alternatively as the facilitator), it is necessary to (briefly) position the main argument of this chapter within the context of ethnicity and tradition. Thus, I begin by briefly examining the question of ethnicity and the problematic 'construction' of ethnic affiliation, including a discussion on the linkages between ethnicity, tradition and cultural/ethnic tourism. Following this, an historical analysis of Carib identity is presented using primary documents and their interpretations by historians and anthropologists. Finally, the chapter attempts an assessment of how specific histories may be at work through cultural tourism development initiatives.

Ethnicity and tradition

Ethnicity is a difficult subject to ignore, yet an equally problematic exercise in recognition. Differentiation between groups or individuals is often fostered regardless of the presence and rise of, as Eriksen (1993) argues, modernisation, industrialisation and individualism. In fact, the later half of the twentieth century has witnessed dramatic growth in the 'culturalisation' of peoples, such that, as Sahlins (1995: 378) points out, ' "Culture" – the word itself, or some local equivalent – is on everyone's lips. Tibetans and Hawaiians, Ojibway, Kwakiutl and Eskimo, Kazakhs and Mongols, native Australians, Balinese, Kashmiries and New Zealand Maori: all now discover they have a "culture"'.

The spread of Western ideological and philosophical elements is primarily responsible for this, suggests Sahlins (1995), and the result has been the reification of cultural differences, distinctive traditions and unique customs, almost to the point where their presence is meant to signify nothing more than overt difference from the dominant, Western 'other'. Sahlins explains:

> More than an expression of 'ethnic identity' – a normal social science notion which manages to impoverish the sense of the movement – this cultural consciousness, as Turner remarks of Kayapo, entails the people's attempt to control their relationships with the dominant society, including control of the technical and political means that up to now have been used to victimize them. The empire strikes back. We are assisting at a spontaneous, worldwide movement of cultural defiance, whose full meanings and historic effects are yet to be determined.
>
> (1995: 379)

Tambiah (1995) suggests that ethnic identities comprise elements of inheritance, ancestry, descent, place of origin, sharing of kinship, skin colour, language, religion, spatio-geographic location, or any combination of these, and arguably many more, attributes. Ethnic identity as a conscious identity is, for Tambiah (p. 430), 'vocalized' in the sense that it 'substantiates and naturalizes' (ibid.) the many elements within an ethnic system used for identification and socialisation purposes. Further, the nature of many modern ethnicities and identities is embedded in a process Tambiah calls 'politicization', such that:

The awareness that collective ethnic identity can be used and manipulated in political action is of course related to the increasing possibilities of contact through the improvement of transportation, of the
quick adoption and deployment of modern media, and of the raised
levels of education and literacy and the spread of what Benedict
Anderson (1983) has called 'print capitalism'.

(p. 436)

Allied closely with the concept of ethnicity is notion of tradition. In one
sense, tradition is not entirely unlike folklore, and in many ways, not
entirely distinct from the broad characteristics that capture the meanings
of ethnicity, as it may more or less represent the social constructions of
events, norms, values, beliefs and understandings by members within and
between groups. It transcends local–global and local–national continuums,
for it can exist within the politico-social organisation of almost any
'society' (however loaded such a term in postcolonial messes may be) or
social group. In his own cautious approach to understanding tradition, Forte
(1998) remarks that, in effect, all tradition is invented, and that any compre-
hension of, for example, aboriginal social organisation necessitates a close
examination of the multiple elements of indigeneity and/or the creation of
indigenous forms and knowledge. In other words, that which constitutes
emic constructions of tradition in relation to identity effectively trumps
colonial-based, positivist exercises of social classification.

On another level, however, tradition can be taken as an overt link to the
historic past which can be manifested in a number of forms. It is this link
with the past through which tradition gains legitimacy, primarily because,
as a social process, it entails historical interpretations that serve to justify
action, beliefs, and, perhaps most importantly, group cohesion. In fact, to
take the term at its face value, it can stand to mean just about anything that
an individual or group of individuals wishes to express. If, however, it is
taken into a historical context, then it is precisely this historical element
which itself governs repetition, therefore allowing for a more 'believable'
and, in one sense, accurate tradition in its own right. In other words, the
difference between repetition and tradition can be, for argument's sake, both
the justification of invention and the degree of innovation. While tradition,
or the practice of tradition, might be seen to have social and cultural justification, underlying historico-social processes are often at work and carry
significant connotations for how such tradition is viewed. In fact, the problematic nature of invented tradition (even all tradition) becomes even clearer
when one examines exactly *which* history is used as a legitimator:

the history which became part of the fund of knowledge or the ideology
of nation, state or movement is not what has actually been preserved
in popular memory, but what has been selected, written, pictured, popu-
larized and institutionalized by those whose function it is to do so.

(Hobsbawm 1983: 13)

Borofsky (1987) puts forward the claim that traditions are continually changing. It is hard to disagree. In one sense, the dynamic nature of traditions within a society can be expressed as a form of discontinuity, to the point where a term such as 're-engineering' can be forwarded (Forte 1996). Such re-engineering represents an alternative conceptualisation of what constitutes ethnicity and tradition. Even more complex and intriguing, however, is when such re-engineering is utilised in the context of postcolonial environments and tourism.

Ethnicity, cultural tradition and tourism: some linkages

That tradition can be seen to be invented, and that ethnicity can exhibit strains of Tambiah's politicisation thesis, has significant consequences for how cultural or ethnic tourism products are developed and presented. This is especially true in postcolonial environments as it is here where negotiation of identity has been most active. Tambiah's politicisation, as a means by which ethnic identities are more or less operationalised for purposes *other* than merely highlighting differences between near neighbours, can ultimately help us decipher and excavate the meaning of a cultural tourism product. In fact, tourism itself can be responsible for what Tambiah has referred to as 'innovations': 'The time of becoming the same is also the time of claiming to be different. The time of modernizing is also the time of inventing tradition as well as traditionalizing innovations; of revaluing old categories and recategorizing new values' (Tambiah 1995: 440).

It is when these 'innovations' are met with certain economic realities that the push for the marketing and selling of ethnic identities to the tourist/traveller becomes almost necessary (Chaney 2002). The relationship between ethnic groups and tourism brings together the quest for the authentic on the part of the tourist and the rationalised and demonstrated ethnicity of the host, especially where economic gain on the latter is both paramount and sought. Often, such collision occurs in postcolonial environments (e.g. Caribbean, Africa) where the colonial army with guns has been all but replaced by the postcolonial army with cameras and guidebooks. What is more, where else to best experience 'the authentic' than in locations where colonialist endeavours sought to initially (and perhaps even successfully) reform and re-task 'native' cultural identities?

Cultural holdovers are irresistible to today's modern cultural tourists. Providing 'the authentic' becomes all too important and can be, proponents would argue, quite positive. The resiliency, however, towards such cultural tourism development is well noted. Hitchcock (1997: 96) refers to at least half a dozen companies on the Internet (there may be more) that market 'Bushman tourism' in the Kalahari Desert. The result was the creation of 'model villages' in Botswana and Namibia which were 'prepared' by governments so local residents can perform 'traditional' activities for the benefit of tourists. One Bushman told Hitchcock (1997: 98): 'We do not want to have to perform for tourists. It is not right that we should be treated like animals in a circus.' This view is emphasised in at least one non-governmental organisation report found by Hitchcock (1997: 102) where it was concluded that 'tourism has done more to promote hunting and gathering than all the efforts of anthropologists put together'. Anthropologists need not feel slighted: participant observation by Hitchcock among the Bushmen revealed that many would run away when the sounds of tourist buses announced their arrival, and one tourist indicated that 'whenever we got close to a Bushman village, we could see people scurrying away' (Hitchcock 1997: 103).

In his study of Bali, Picard (1992) found that the country itself had become a 'living museum', so much so that Picard uses the word 'touristification' (p. 60) to describe a situation in which it is no longer possible, in his view, to separate that which is entirely indigenous from that which has been effectively influenced by tourism development. Moreover, Wall (1998: 55) noted that the official policy for Bali is '*pariwisata budaya*', or cultural tourism, despite the fact that cultural tourism more or less falls to the bottom of the trip mix of many tourists due to the high demand for sightseeing holidays.

The Carib of St Vincent – constructing histories and ethnicities

The small island-state of St Vincent and the Grenadines lies approximately 160 kilometres to the west of Barbados and some 320 kilometres from the northern coast of South America (see Figures 4.1 and 4.2). Compared to other island states in the Eastern Caribbean, St Vincent and the Grenadines' entry into the regional tourism market has been relatively recent (late 1980s and early 1990s). A considerable portion of its gross domestic product is dominated by agriculture, although tourism is increasingly being positioned as a key source of economic activity. As a result, the government of St Vincent and the Grenadines has been actively planning (in order to compete with neighbouring islands such as Barbados, St Lucia and Grenada) for the expansion of the local tourism sector.

St Vincent's precarious position in the southern Caribbean Sea, close to more intensively developed islands (including tourism) such as Barbados and St Lucia, has been the most significant attraction for tourists from Europe, Canada and the United States. Unlike Barbados and St Lucia, however, St Vincent has limited facilities and services to offer the some 20 million visitors who visit the region each year. In 2002, some 70,000 tourists visited St Vincent and the Grenadines (Caribbean Tourism Organization 2003). Interestingly, St Vincent's immediate competition can be found to the south in the neighbouring, and politically ceded, Grenadine islands, instead of Barbados and St Lucia. As a result, the government of St Vincent and the Grenadines actively promotes a tourism programme that can best be described as one based on alternative tourism. In the absence of conventional mass tourism on St Vincent, alternative forms of tourism may be more feasible, although perhaps not as financially rewarding due to constrained capacities, limited marketing budgets and variable (and fickle) demand (Duval 1998).

In the early 1990s, communities in the north of St Vincent were involved in exercises designed to promote 'nature tourism' (which included cultural representations) through a variety of community-based development projects (Franklin and Morley 1992; Franklin 1993). At the time, pressures on banana production in St Vincent were mounting. The preferential market for Eastern Caribbean bananas was under threat by US-led challenges of the existing Lomé Conventions for such arrangements. Tourism, especially in remote, rural areas of the Caribbean (a prime location for all those 'new' tourists looking for the alternative experience) seemed suitable to pick up the slack left by a crumbling agricultural sector (Payne and Sutton 2001). At the time, Barbados, Antigua and many other islands in the region were already reaping the economic benefits of tourism (in theory), so many of the smaller, agriculturally focused islands of the Eastern Caribbean (e.g. St Vincent, Dominica) actively considered bolstering their tourism sector. Such is the context in which the 'nature tourism' development exercises, described below, can be situated.

A team of researchers from Canada initiated a six-year, four-island project (St Vincent, Grenada, Dominica and St Lucia), funded by the Canadian International Development Agency, designed to promote nature tourism through community-based initiatives. These initiatives were designed to compensate for fluctuations and 'cynicism' associated with governmental development aid (Franklin 1993: 6). The project sought to use nature tourism 'as the context for examining the tension associated with balancing economic growth, environmental conservation, and cultural sustainability' (Franklin 1993: 4).

In 1991 and 1992, National Advisory Groups were formed in each country under the direction and guidance of Franklin and Morley. Each group consisted of representatives from specific government agencies (forestry, national parks, tourism), various NGOs (non-governmental organisations) – National Trusts and existing environmental organisations, individuals from the business community. Next, a series of contextural 'search conferences' were initiated, with the Groups ultimately controlling their direction and scope (Franklin and Morley 1992: 239). As a result, grassroots initiatives toward development planning and management, and involving local community participants (through the process of 'action learning'), were established. Of interest here is how the 'nature tourism' project played out on St Vincent, and especially in the context of the Carib community, who live along the northern coast of the island, where a particular component of the project was centred (Franklin 1998: 188–93). Outcomes of the project included 'restoration of an historical Carib well

Figure 4.1 The Eastern Caribbean

site and the construction of signs announcing the various villages that make
up the Carib community' (Franklin 1998: 189). Taking on a role roughly
equivalent to Fagence's (2001) 'custodians', the project's national coordinator for St Vincent situated the project as follows:

> This project reflects an attempt on the part of the people of St Vincent's
> Carib community to strengthen their recognition as an indigenous
> people. The Carib community leaders are attempting to develop possi-
> bilities for economic and physical improvement. While there is a phys-
> ical infrastructure component to the project, the main purpose is to
> open up economic and cultural possibilities for the Carib people.
>
> (quoted in Franklin 1993: 8)

Providing or making available the tools necessary for such grassroots initiatives is becoming more common in the context of tourism development
within developing countries. With this emphasis on cultural values, however, comes the need for recognition of existing power structures (Bianchi
2003). This particular project focused on the development of tourism as
a tool for economic growth, with one of the overall objectives being 'to
provide an opportunity to treat nature tourism as a common development

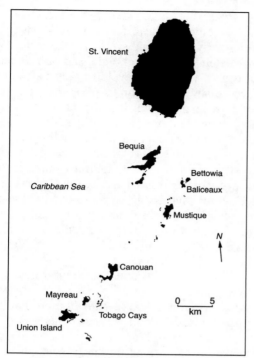

St. Vincent

Bequia

Bettowia

Caribbean Sea

Baliceaux

Mustique

N

Canouan

Mayreau

0   5
km

Tobago Cays

Union Island

Figure 4.2 St Vincent and the Grenadines

issue among the four Windward Island nations' (Franklin 1998: 143), and in one sense, responds to Vincent's charge that, at least in the Caribbean, community-based tourism initiatives are few and far between:

> There is still not enough research on the degree to which local people really feel they have some input into and control over the future development of the industry. Such research should define ways in which participation could be maximized – a vital factor in creating an industry that meets the needs and desires of the majority of the local population.
>
> (1995: 261)

Outside of the viability of the research design forwarded by the project (which is not under scrutiny here), and irrespective of the beneficial aspect of involving local participation in development planning at the community level, significant socio-cultural implications, many with substantial colonial overtones, come into play with respect to the results of the research. These may not have been initially realised by the researchers and participants. Consider, for example, the following quotation from St Vincent's national coordinator for the project:

The project has two goals: firstly, the clearer identification of the Carib community through graphic labelling of the different villages – attempting to give them an identity for visitors; and secondly, the restoration of an old well that the indigenous people used to depend on for water. The Caribs want to reconstruct the well and develop a picnic area close to it where community people and visitors might enjoy the sight. The well-site represents a centre of their culture; it is the site of the original village and represents a tangible expression of their history. They imagine using it on moonlit nights for communal story telling.

The project is part of a wider process. For example, they also plan to convert an old police station for use as a cultural centre, so that visitors can experience displays of performing arts by the Carib people. Another aspect of the long-term plan is the carrying out of research and the promotion of historical information about the community. They want to publish a brochure on historical events and the sites in the community and to produce a video involving the entire St Vincent community as well as the Caribs in Dominica.

(quoted in Franklin 1993: 8)

The project, seemingly designed to both preserve and promote national heritage and development, may nonetheless initiate and foster a specific cultural heritage for the purposes of tourism. Franklin even makes mention of various ideas regarding potential extensions of the project, many of which take on a tone relating to future tourism development:

Many of these had to do with Carib identity: developing their own historical knowledge by setting up a cultural museum in the old police station at Sandy Bay, retrieving Carib pottery and artifacts that had been taken from the community by collectors, and relearning old crafts such as basketwork and pottery making.

(1998: 191)

It will be argued below that what is being developed is essentially *a* history, and not *the* history, of the St Vincent Carib. Despite the assertion made by the national coordinator above, the Carib of St Vincent are somewhat distinct from the Carib of Dominica (see Layng 1983), and the meaning of the term 'Carib', with reference to St Vincent specifically, needs further contextualisation from a historic and ethnohistoric perspective.

The ethnohistory of the Carib

In providing a brief ethnohistory of the Carib, my intent is to draw out the intersection of histories, ethnicities and cultural tourism discussed above. For our purposes, at issue is *which* 'Carib' history is being put forward. To this end, the national coordinator's comments above regarding the promotion of the historical aspect of the Carib villages are somewhat intriguing in that the true history of the Island Carib on St Vincent, despite decades of historiographic research, still remains somewhat of a mystery.

On the surface, members of the Carib community of St Vincent trace their origins and cultural ancestry back to a culture group broadly known as the 'Carib', yet in fact there are at least two distinct groups to consider in the ethnohistory of indigenous St Vincent: the Black Carib and the Island Carib, the latter of which, to confuse matters even more, are often referred to in primary historical documents as either the Red Carib or Yellow Carib. (For the sake of simplicity, the term 'Island Carib' shall here incorporate these alternative 'Red' or 'Yellow' characterisations.) Overall, recognition of the existence of this Island/Black Carib dichotomy (characterisations generally accepted among ethnohistorians, archaeologists and anthropologists) goes some way to address difficult and confusing ethnohistorical problems with reference to the notions of ethnicities and histories.

The pre-historic/historic Island Carib have long been an elusive ethnic group. They likely arrived in the Lesser Antilles from the north coast of South America shortly before Columbus (Allaire 1980). They are associated with the first European settlers (Columbus is even said to have met some) and generally occupied the Windward Islands during the early historical period (Allaire and Duval 1995). Historical accounts of the Island Carib are scarce. Commissioned (an important point to remember) and freelance British and French writers of the early seventeenth to late eighteenth centuries say relatively little about the Island Carib, although Father Breton's *Dictionnaire caraïbe-français* (1892 (1665)), provides significant insight into their social customs and traditions.

The absolute genesis of the Black Carib is somewhat more murky than the Island Carib. It is well known that black slaves often escaped from the clutches of European oppression throughout the islands during the sugar revolution of the seventeenth century (Du Terte 1667; Gonzalez 1988). As a result, they would often seek refuge in the dense forests of less developed islands such as St Vincent. Over time, these runaway black slaves mixed with some Island Carib populations on the island. Another origin myth is that a slave ship was wrecked off the southern coast of St Vincent, and the occupants waded ashore and began to live with the Island Carib. Regardless of their precise origins, what is important here is the genetic mixing that took place between black slaves and Island Carib to produce a new ethnic group. Thomas Coke's commissioned *A History of the West Indies* (1971 (1810): 180) suggests that, over time, the Black Carib increased in number, acquiring 'power' and particularly ambition. Young (1993: 23) has suggested that the Black Carib came to be known to the French as the 'Black Carib' by at least 1700.

In *An Historical Account of the Island of Saint Vincent,* nineteenth-century historian Charles Shephard remarked that, having distinguished themselves from the incoming black slaves used by the French, the Black

Carib of St Vincent continued trade with the French and established themselves as a separate 'tribe' (Shephard 1971 (1831): 24). Politically, they were organised by a chief, and continued to engage in warfare against the Island Carib. Shephard (1971 (1831): 24–5) suggested that both Black Carib and Island Carib may have co-existed on St Vincent for quite some time. This reference to a separate Black Carib identity on St Vincent is further supported by Coke's indication of the reaction on the part of the Black Carib when French planters came to settle on St Vincent, bringing with them indentured Negro slaves:

> The black Charaibs, shocked at the idea of resembling men who were degraded by slavery, and fearing that in process of time their own color, which betrayed their origin, might be made a pretense for enslaving them, took refuge in the thickest part of the woods. In this situation, in order to create and perpetuate a visible distinction between their race and the slaves brought into the island, they compressed so as to flatten the foreheads of all their new-born infants; and this was, thereafter, considered as a token of their independence. The next generation thus became, as it were, a new race.
>
> (Coke 1971 (1810): 181f)

The Black Carib (often called, simply, the Carib by colonial settlers and military forces) flourished until their eventual deportation from the island by British troops in the late eighteenth century following numerous armed skirmishes (the great 'Carib Wars' as described in primary historical documents) and the rampant spreading of disease (Gonzalez 1988). The British contended at the time that all 'Carib' were eradicated or were forcibly removed. The Black Carib descendants, known as the Garifuna, now live in Belize, Honduras and the tiny island of Roatan off the Central American coast. What is not entirely clear, however, is what became of the Island Carib. The actual numbers of Island Carib residing in St Vincent at the end of the eighteenth century remain in doubt. The historian Thomas Atwood, writing in the late eighteenth century (Atwood 1971 (1791)), hinted that 'there are not more than twenty or thirty families' left on Dominica, and Kipple and Ornelas (1996) postulate that these may even have been the last remaining Island Carib in the region. Most primary sources suggest that they were incorporated into the Black Carib population, and that by the time of their forced removal to Central America, the only 'Carib' on St Vincent were in fact Black Carib.

The above is not meant to be a thorough ethnohistorical sketch of St Vincent, but it is meant to convey the fact that St Vincent has played host to an intricate series of events in its culture history, many of which historians still wrestle with in terms of accuracy. As well, *the* culture history of St Vincent is far from certain, open to numerous interpretations, and subject to sceptical use of commissioned colonial reports. In one sense, the formal identification of the Black Carib by Europeans was somewhat unstructured and leads us to consider seriously the accuracy of modern representations of a Carib identity, as many early writers would often simply refer to the Black Carib as the Carib. That being said, what is important to remember is that reference to modern Carib identity on St Vincent invites more questions than it answers with respect to social history and representation.

While St Vincent witnessed a significant amount of mixing between Island Carib and Black slaves, the same cannot be said for Island Carib of Dominica. In fact, several families, descendants of Island Carib, currently reside on the island of Dominica, where the government established a Carib Reserve in the northern portion of the island in the 1950s (Layng 1983), and they themselves have been active in the development of culture-based tourism (Slinger 2001). Even in Dominica, one may question those elements involved in defining typical Carib lifestyle (very few attempts have been made, but some exceptions do occur (Layng 1983; Gullick 1995)); but of greater interest is how it is being presented for the purposes of increasing local revenue from tourism. At least one attempt has been made to capture the ascribed (and even self-ascribed) identity of the Carib on Dominica:

> People here not really Caribs; many have straight hair, but all are mixed. I'm not a Carib, but all people here are called Carib. A few older ones, but their children have mixed, so no more Caribs. Even straight hair does not mean real Carib.
> (older basketmaker from Gaulette River in Dominica, quoted in Layng 1983: 144)

Further, Baker reasons that the cultural characteristics of the Carib on Dominica are no longer what they once were:

> The Caribs define themselves as Carib today in the absence of almost all tradition Carib cultural criteria. The language of the Carib is the local French *patois*, their religion is Catholic, their dwellings are typical of the other peasants on the island, they own, and are proud of the productivity of, their banana trade.
> (1988: 392)

Lowenthal (1972; see also Rae and Armitage-Smith 1932), in a discussion surrounding the creolisation of ethnic minorities in the Caribbean, emphasises that Lesser Antillean Carib are 'almost completely' creolised:

> Basket-making (promoted by Social Welfare Officers) and inheritance through the male line are the only 'Indian' traits of the Dominica Caribs. The Dominica Carib Reserve, along with the Carib Chief and Council, are early twentieth-century creations. The Governor who formalized

the Carib Reserve tried 'to make them realize that they are the last remnant of a fine race and that should try to keep their breed pure', but a generation later a Commission reported that 'they have no folklore, no songs or music, no dances or customs, no costume or ornament to distinguish them from the other inhabitants of Dominica'.

(1972: 182)

Lowenthal's comments follow from Owen's (1975) analysis, which states that Carib ethnicity in Dominica is an adaptive strategy as opposed to a 'product of cultural conservation or isolation'. Government-sanctioned identity is often the strongest kind, and as Picard (1992) notes for Bali, often quite contested.

For Dominica, then, questions of Carib ethnicity have already been raised. For St Vincent, where the culture history/ethnohistory is more complex, can we even identify and expect to comprehend a modern manifestation of 'Carib' ethnicity? The question, however, is not really whether or not the particular form of cultural tourism development taking place in St Vincent can be considered to be based upon authentic 'Carib' lifestyles, but rather how that lifestyle, as a form of cultural tourism, is constructed. What exactly is being presented? More precisely, can the Carib of modern St Vincent be considered descendants of Island Carib or Black Carib?

One history would suggest that Carib did in fact prosper in the northwest coast of St Vincent. Another would suggest that, based on ethnohistorical research, the ethnic moniker of 'Carib' is confusing, inaccurate and subject to colonialist and loosely sponsored accounts of defeat. In light of the ethnohistorical evidence that provides a closer view of issues of disease, warfare, social and cultural intermixing, perhaps what we have is a modern social group who claim a historic, postcolonial lineage of social association(s) engaging in what Dogan (1989) refers to as a revitalisation of culture. As Bendix notes:

> Cultural displays require staging and thus negotiation of some sort; even a rite of passage is newly created by active participants who decide how and when the event is to take place in ever-changing social conditions. Tourism and its concerns simply add a further element in the staging process. In conjunction with a tourist economy, it is then precisely the realm of expressive culture and its strategic use by the host society that allows for a more differentiated analysis of tourism's impact on the host's culture, or rather, the degree of cultural resilience on the part of the hosts in the face of tourism.

(1989: 143)

Assessment and conclusion

The above example extends our consideration of authenticity by (1) encouraging consideration of specific histories that dictate and frame what is authentic and what is not from the perspective of production, and (2) suggesting that the nature of community-based cultural tourism development incorporates specific histories that can often mask historical incongruities and debates. In bridging the ethnohistory of the Carib with the 'nature tourism' project initiated by Franklin and Morley, what is evident with respect to the 'Carib' is that specific histories intersect with cultural traditions and, in this case, the drive for the development of tourism that itself is sponsored by foreign intervention on the part of researchers funded by a Western aid agency. But it is not entirely clear whether the contested histories presented above are even considered by the actors, nor whether tourists will even consider (or care about) these significant historical questions. As a result, key historical interpretations are absent in the community-based project initiated by Franklin and Morley, despite the fact that 'this alternative form of tourism is based on the attraction of natural, cultural, and historical environments and attempts to balance the economic objectives of tourism with environmental conservation and cultural/historical preservation' (Franklin and Morley 1992: 239). What this project perhaps best demonstrates, however, is not an exercise in cultural or historical preservation, but rather an exercise in which the sacred is constructed precisely in order for the profane to be created.

Obfuscated within this discussion, however, has been the extent to which authenticity is created and developed. Marwick (2001: 48), for example, points to the production of handicrafts in Malta as almost a delicate balance between 'complementary commercialization' and 'substitutive commercialization'. In other words, some forms of 'cultural production', whether these are performances or tangible goods such as souvenirs (or both), may be produced in great quantities in order to 'satisfy the mass of tourists' (Marwick 2001: 48). Lacy and Douglass (2002) suggest that some tourism performances might well reveal more of the processual nature (Olsen 2002) of identity than would first be implied: 'What are being configured and then promoted, of course, are various "versions" of Basque identity. Indeed, we would argue that tourism's propents and opponents alike have vested interests in a (i.e. their) determination of Basque cultural authenticity' (Lacy and Douglass 2002: 17–18). To revisit Tambiah, producing culture for touristic consumption is often highly politicised. Indeed, Hall (1998: 140) remarks that 'the process of "producing" cultural landscapes for tourism consumption makes one dependent on the other, for there can be no consumption without production'. Hall's point echoes that of Mowforth and Munt (1998), who recognise that, as trends in tourism consumption change, so will the production/offerings of tourist experiences. In fact, one of the purposes of tourism, if we follow MacCannell (1976), is often to promote that which is not real, yet at the same time, and for political and socio-economic reasons, that which is more or less economically justified.

This chapter, then, has not been concerned with the individualised and contested context of the consumption of authenticity (e.g. Olsen 2002), nor has it made an attempt to contribute to the lively debate of how authenticity is experienced (Wang 2000) and whether tourists, in a Marxist sense, consume the Other. Rather, it has taken a different approach and examined authenticity from the standpoint of production, and runs almost contrary to Fagence's (2001: 10) arguments that, in some cases, 'strategic intervention' in the management of cultural tourism can afford a degree of protectionism of cultural values. What has been considered here is not the implications and processes necessary for managing cultural tourism in the many postcolonial environments where identities are mixed, muted and dynamic, but rather whether the type of intervention, as outlined by Fagence (2001), necessarily leads to the reification of cultural elements for the purposes of touristic voyeurism. In other words, what is integral is 'how authenticity is constructed and gets decided' (Shephard 2002: 19.6). In effect, what has been argued above is that, when tradition and identity are invented, the differences between the sacred and the profane become moot, just as the argument for authentic/inauthentic does little to capture the process or meaning behind cultural tourism.

In the case of the Carib, by commodifying an invented (or, at the very least, a historically problematic) tradition for the purposes of tourism, and by generating what may or may not be considered authentic based on constructed cultural associations, the Carib may be allowing for the creation of a distinct group that provides an ideal example for the anthropological designation of the invention of tradition. This is, in one sense, beyond distinctions between the sacred and the profane, the separation and delineation of which Shephard (2002) argues is diluted enough such that 'local actors' cannot clearly delineate between the two.

Acknowledgements

Thanks to Tim Coles for, as always, exceptionally valuable comments on an early draft. Thanks also to Eric Shelton for his own insightful comments and suggestions.

References

Allaire, L. (1980) 'On the historicity of Carib migrations in the Lesser Antilles', *American Antiquity* 45: 238–45.

Allaire, L. and Duval, D. T. (1995) 'St Vincent revisited', in R. E. Alegria and M. Rodriguez (eds) *Proceedings of the XV International Congress for Caribbean Archaeology*, Centro de Estudios Avasnzados de Puerto Rico y el Caribe, con la colaboracion de la Fundacion de las Humanidades y la Universidad del Turabo: San Juan, Puerto Rico, pp. 255–62.

Anderson, B. (1983) *Imagined Communities: Reflections on the Origin and Spread of Nationalism*, London: Verso.

Atwood, T. (1971 (1791)) *The History of the Island of Dominica. Containing a description of its situation, extent, climate, mountains, rivers, natural productions*, London: Frank Cass.

Baker, P. L. (1988) 'Ethnogenesis: the case of the Dominica Caribs', *América Indígena* 48: 377–401.

Bendix, R. (1989) 'Tourism and cultural displays: inventing traditions for whom?', *Journal of American Folklore* 102: 131–46.

Bianchi, R. (2003) 'Place and power in tourism development: tracing the complex articulations of community and locality', *PASOS: Revista de Turismo y Patrimonio Cultural* 1, 1: 13–32.

Boorstin, D. (1964) *The Image: A guide to pseudo-events in America*, New York: Harper and Row.

Borofsky, R. (1987) *Making History: Pukapukan and anthropological constructions of knowledge*, Cambridge: Cambridge University Press.

Breton, R. P. R. (1892 (1665)), *Dictionnaire caraïbe-français*, Leipzig: B. G. Teubner.

Caribbean Tourism Organization (2003) *Caribbean Tourism Statistical Report, 2001–2002 edition*, Barbados: Caribbean Tourism Organization.

Chaney, D. (2002) 'The power of metaphors in tourism theory', in S. Colemand and M. Crang (eds) *Tourism: between place and performance*, New York: Berghahn Books, pp. 193–206.

Cohen, E. (1995) 'Contemporary tourism – trends and challenges: sustainable authenticity or contrived post-modernity?', in R. Butler and D. Pearce (eds) *Change in Tourism: People, Places, Processes*, London: Routledge, pp. 12–29.

Coke, T. (1971 (1810)) *A History of the West Indies containing the Natural, Civil, and Ecclesiastical History of each Island: with An Account of the Missions Instituted in those Islands, from the Commencement of their Civilization; but more especially of the Missions which have been Established in that Archipelago by the Society Late in Connection with the Rev. John Wesley*, London: Frank Cass.

Dogan, H. (1989) 'Forms of adjustment: sociocultural impacts of tourism', *Annals of Tourism Research* 16: 213–36.

Du Terte, J. B. (F.) (1667) 'Concerning the natives of the Antilles', extract from *Histoire Generale des Isles de S. Cristophe, de la Guadeloupe, de la Martinique, et autres dans l'Amerique*, trans. M. McKusick and P. Verin, Human Relation Area Files ST13 M-4 (1958).

Duval, D. T. (1998) 'Alternative tourism on St Vincent, West Indies', *Caribbean Geography* 9, 1: 44–57.

Edensor, T. (1998) *Tourists at the Taj: Performance and meaning at a symbolic site*, London: Routledge.

Edensor, T. (2001) 'Performing tourism, staging tourism: (re)producing tourist space and practice', *Tourist Studies* 1, 1: 59–81.

Eriksen, T.H. (1993) *Ethnicity and Nationalism: Anthropological perspectives*, London: Pluto Press.

Fagence, M. (2001) 'Cultural tourism: strategic interventions to sustain a minority culture', *The Journal of Tourism Studies* 12: 10–21.

Forster, J. (1964) 'The sociological consequences of tourism', *International Journal of Comparative Sociology* 5: 217–27.

Forte, M.C. (1996) 'Of blood and names and colonial ancestry: the re-engineering of Carib indigeneity in Trinidad and Tobago', Paper presented at the 27th Annual Congress of the Canadian Association for Latin American and Caribbean Studies, 31 October to 3 November, 1996, York University, Toronto, Canada.

Forte, M.C. (1998) 'The international indigene: regional and global integration of amerindian communities in the Caribbean', Paper presented at the 1998 Congress of the Canadian Association for Latin American and Canadian Studies, 19–21 March, 1998, Simon Fraser University, Vancouver, Canada.

Franklin, B. (1993) 'Grassroots initiatives in Sustainability: A Caribbean example', in D. Bell and R. Keil (eds) *Human Society and the Natural World: perspectives on sustainable futures*, Toronto: Faculty of Environmental Studies, York University, pp. 1–10.

Franklin, B. (1998) 'Contextual action research: extending praxis methodology', PhD dissertation, The Fielding Institute.

Franklin, B. and D. Morley (1992) 'Contextural searching: cases from waste management, nature tourism, and personal support', in M. Weisbord (ed.) *Discovering Common Ground: How Future Search Conferences Bring People Together to Achieve Breakthrough Innovation, Empowerment, Shared Vision, and Collaborative Action*, San Francisco: Berrett-Koehler, pp. 229–46.

Gonzalez, N. L. (1988) *Sojourners of the Caribbean: Ethnogenesis and Ethnohistory of the Garifuna*, Urbana: University of Illinois Press.

Gullick, C. J. M. R. C. (1995) 'Communicating Caribness', in N. L. Whitehead (ed.) *Wolves from the Sea: Readings in the Anthropology of the Native Caribbean*, Leiden: KITLV Press, pp. 157–70.

Hall, C. M. (1998) 'Making the Pacific: globalization, modernity and myth', in G. Ringer (ed.) *Destinations: Cultural Landscapes of Tourism*, London: Routledge, pp. 140–53.

Hitchcock, R.K. (1997) 'Cultural, economic, and environmental impacts of tourism among Kalahari Bushmen', in E. Chambers (ed.) *Tourism and Culture: An Applied Perspective*, New York: State University of New York Press, pp. 93–128.

Hobsbawm, E. (1983) 'Introduction: inventing traditions', in E. Hobsbawm and T. Ranger (eds) *The Invention of Tradition*, Cambridge: Cambridge University Press, pp. 1–14.

Hughes, G. (1995) 'Authenticity in tourism', *Annals of Tourism Research* 22, 4: 781–803.

Kipple, K. F. and Ornelas, K. C. (1996) 'After the encounter: disease and demographics in the Lesser Antilles', in R. L. Paquette and S. L. Engerman (eds) *The Lesser Antilles in the Age of European Expansion*, Gainesville, FL: University Press of Florida, pp. 50–70.

Lacy, J. A. and Douglass, W. A. (2002) 'Beyond authenticity: the meanings and uses of cultural tourism', *Tourist Studies* 2, 1: 5–21.

Layng, A. (1983) *The Carib Reserve: Identity and Security in the West Indies*, Washington, DC: University Press of America.

Lowenthal, D. (1972) *West Indian Societies*, London: Oxford University Press.

MacCannell, D. (1976) *The Tourist: A New Theory of the Leisure Class*, New York: Schocken.

Marwick, M.C. (2001) 'Tourism and the development of handicraft production in the Maltese islands', *Tourism Geographies* 3, 1: 29–51.

McKercher, B. (2002) 'Towards a classification of cultural tourists', *International Journal of Tourism Research* 4: 29–38.

Mowforth, M. and Munt, I. (1998) Tourism and Sustainability: New Tourism in the Third World, London: Routledge.

Olsen, K. (2002) 'Authenticity as a concept in tourism research: the social organization of the experience of authenticity', *Tourist Studies* 2, 2: 159–82.

Owen, N. H. (1975) 'Land, politics and ethnicity in a Carib Indian community', *Ethnology* 14, 4: 385–94.

Payne, A. and Sutton, P. (2001) *Charting Caribbean Development*, London: Macmillan.

Picard, M. (1992) Bali: Tourisme culturel et culture touristique. Paris: L. Harmattan.

Rae, J. S. and Armitage-Smith, S. A. (1932) 'Report of a committee appointed by His Excellency the Governor of the Leeward Islands to enquire into conditions in the Carib Reserve, Dominica, and the disturbance of 19th September, 1930', London: HMSO, Cmd 3990.

Sahlins, M. (1995) 'Goodbye to tristes tropes: ethnography in the context of modern world history', in R. Borofsky (ed.) *Assessing Cultural Anthropology*, 2nd edn, New York: McGraw-Hill, pp. 377–95.

Shephard, C. (1971 [1831]) *An historical account of the island of Saint Vincent*, London: Frank Cass.

Shepherd, R. (2002) 'Commodification, culture and tourism', *Tourist Studies* 2, 2:183–201.

Slinger, V. (2001) 'Ecotourism in the last indigenous Caribbean community', *Annals of Tourism Research* 27, 2: 520–23.

Tambiah, S. J. (1995) 'The politics of ethnicity', in R. Borofsky (ed.) *Assessing Cultural Anthropology*, 2nd edn, New York: McGraw-Hill, pp. 430–41.

Titley, G. (2000) 'Global theory and touristic encounters', *Irish Communications Review* 8: 79–87.

Tucker, H. (2001) 'Tourists and troglodytes: negotiating for sustainability', *Annals of Tourism Research* 28, 4: 868–91.

Urry, J. (1990) *The Tourist Gaze: Leisure and Travel in Contemporary Societies*, London: Sage.

van den Berghe, P. (1994) *The Quest for the Other: Ethnic Tourism in San Cristobal, Mexico*, Seattle, WA: University of Washington Press.

Vincent, G. (1995) 'Tourism and sustainable development in Grenada, West Indies: towards a mode of analysis', PhD Dissertation, Department of Geography, McGill University, Montreal.

Vivanco, L. A. (2002) 'Ancestral homes: indigenous peoples are pushing for tourism alternatives that respect community, culture and the land', *Alternatives Journal* 28, 4: 27–8.

Wall, G. (1998) 'Landscape resources, tourism and landscape change in Bali, Indonesia', in G. Ringer (ed.) *Destinations: Cultural Landscapes of Tourism*, London: Routledge, pp. 51–62.

Wang, N. (2000) *Tourism and modernity*, Amsterdam: Pergamon.

Young, V. H. (1993) *Becoming West Indian: Culture, Self and Nation in St Vincent*, Washington, DC: Smithsonian Institution Press.

5　About romance and reality

Popular European imagery in postcolonial tourism in southern Africa

Harry Wels

Introduction

The idea of (southern) Africa in Europe has always been dominated by images of landscape and physical aestheticism derived from Romanticism (Grove 1987). 'Aesthetics refers to the philosophical investigation of beauty and the perception of beauty . . .' (Rojek 1995: 165). European images of Africa had to function as contrast to and measure of European 'civilization' (Corbey 1989: 40), being socially constructed ideas about landscape and the people. Rojek (1997: 53) proposes that 'myth and fantasy play an unusually large role in the social construction of *all* travel and tourist sights'. This chapter will discuss some specific examples of the colonial myths and fantasies that shaped the European social constructions of African landscapes and peoples. Furthermore, it will be shown how these colonial images continue to play a powerful role in shaping the current gaze of Europeans in postcolonial tourism.

In the European 'idea of Africa' (Mudimbe 1994), landscapes often come first. African people have to blend in as part of the landscape (Wilmsen 1995). Kirkaldy describes the relation between European missionaries and the landscape of Vendaland in South Africa as:

> what remained constant in missionary thinking and writing in Vendaland during the late nineteenth century was a sense of the land and its people as inextricably bound. (. . .) (T)he 'heathen Bawenda [Vhavenda]', as the missionaries called them, were portrayed as blending into, or being created by, the landscape which nurtured, succoured and concealed them.
>
> (2003: 173)

In similar vein, Angela Impey, a researcher at the University of Natal's School of Music, is quoted in *The Natal Witness* (Anon. 2003) as saying that 'Culture is as integral a part of the treasure of the South African landscape as its fauna, floral and marine resources'. In other words, Africans and African culture should blend into an aesthetically dominated European image of African landscape. That is why European tourists visiting post-colonial Africa nowadays usually perceive huts with thatched roofs and African women with water buckets on their head as 'authentic Africa', while Cape Town is considered to be 'not the real Africa'. Huts and women with buckets on their head fit with European perceptions of African landscapes, while buzzing, cosmopolitan city life is alien to that image.

European imagery of Africa has often been (re)presented, in the most literal sense of the word, in photographs and art. Imagery lies at the basis of an 'us' and 'them' categorisation, which implies the serious risk of leading to 'us' stereotyping the (African) Other (see Pickering 2001). The central issue for reflection in this chapter is the historical contextualisation and discourse of European images of African landscape and culture and its implications for present-day cultural tourism in southern Africa. In order to develop the argument, I will use chronological examples of described imagery in anthropological and other literature with regard to Africa in general and African people in particular (Corbey 1989, 1993; Price 1989; Gordon 1999, 2002; see also Doy 1998; Jahoda 1999; Maxwell 2000; Landau and Kaspin 2002; Lübbren and Crouch 2003).

African landscapes and African Others in European imagery

The concept of 'landscape' has two distinct but related usages. In the first place it denotes 'an artistic and literary representation of the visible world, a way of experiencing and expressing feelings towards the external world, natural and man-made, an articulation of a human relationship with it' (Cosgrove 1984: 9). The second usage is found in current geography and environmental studies: 'Here it denotes the integration of natural and human phenomena which can be empirically verified and analysed by the methods of scientific enquiry over a delimited portion of the earth's surface' (p. 9). These two usages are 'intimately connected both historically and in terms of a common way of appropriating the world through the objectivity accorded to the faculty of sight and its related technique of pictorial representation' (p. 270). It is important to note explicitly that 'the concept of landscape is a controlling composition of the land *rather than its mirror*' (p. 270). So landscape is about constructing images, and its representations 'cannot be divorced from representations of landscapes' inhabitants' (Wolmer forthcoming: note 27). However, the landscape seems always to come first.

In southern Africa, the Scottish Reverend John Croumbie Brown is described by Grove (1997: 140) as the 'single most influential voice' in creating a colonial discourse on landscape. Grove argues that the Scottish landscape and environmental sensibilities were the 'major vehicle' for the expression of their national identity in opposition to the English and English rule. The Scots' perspective was a highly aesthetic one, rooted in

Romanticism, which was firmly wrapped and strengthened by a mythology about their specific Scottish history, which separated and distinguished them sharply from the English. They translated their love for their own Scottish aesthetic landscapes to Africa in general and, for instance, closely followed the Scotsman Mungo Park on his travels through the African interior around the River Niger through newspaper reports. They bewailed his tragic death and made him a martyr whose example should be followed by other Scotsmen. Africa hence became a 'national obsession' in Scotland, according to Grove. Southern Africa was given a starring role because a particular journal, *Penny Magazine*, paid an unusually generous amount of attention to southern Africa through the efforts of Thomas Pringle. This poet wrote numerous articles about the region in which he compared the Scottish landscape, so important to his own and the Scots' sense of social and national identity, with that of South Africa. From there it was only a small step to romanticising and sacralising the landscape in South Africa in the same fashion and deriving a new and strong identity from it (Grove 1997: 142–3). Draper (2003) also uses a Scotsman, James Henderson, but now with a passion for angling and fly-fishing, as an introduction to analyse the relationship between South African landscape and processes of social identity construction of white settlers (see also about the relationship between landscape and national (Afrikaner) identity processes with regard to the development of South Africa's tourist flagship Kruger National Park (Carruthers 1995)). From this perspective it is not difficult to see why the European imagery about Africa's landscape is often expressed in terms of the ultimate aesthetic natural icon, their 'lost Eden'. Draper and Maré maintain that '(n)ationalism thrives on romanticism, not least romanticism about nature' (Draper and Maré 2003: 559). Romanticism had and still has a profound influence on European thinking and imagery (see for example Lemaire 2002). In Anderson and Grove's words:

> Much of the emotional as distinct from the economic investment which Europe made in Africa has manifested itself in a wish to protect the natural environment as a special kind of 'Eden', for the purposes of the European psyche rather than as a complex and changing environment in which people actually have to live . . . (thus) Africa has been portrayed as offering the opportunity to experience a wild and natural environment which was no longer available in the domesticated landscapes of Europe.
>
> (1987: 4)

Africa (or Eden) became synonymous with a European sense of authenticity concerning both nature and the way that people should relate to and blend into nature. 'The game reserve might be said to theatricalize a framing, primal past for modernizing Europe' (Bunn 1999: 9). The usage of the term 'Eden' in describing African landscapes from a European perspective is probably one of the stronger metaphors to describe the norms for perfect natural aesthetics:

> The Edenic vision of the landscape was capable of accommodating an African presence, because incorporated in the Eden myth is the myth of the noble savage. The noble savage, being closer to nature than civilization, could, hypothetically, be protected *as a vital part* of the natural landscape.
>
> (Neumann 1998: 18)

This positive image of an African Eden contrasts sharply with the Europeans' simultaneous fear of the 'dark continent' with all its connotations of death and destruction. Both images taken together – let us call them Eden and Armageddon to continue the Biblical metaphors – form the paradoxical European imagery of Africa(ns): '(o)ne "unresolved ambivalence" has been the incompatible European images of Africa as forbidding wasteland or Edenic paradise' (Rojek 1995: 17). It is not without meaning in this respect that in Joseph Conrad's *Heart of Darkness* the gloomy descriptions of landscape and the stereotypical descriptions of the 'primitiveness' of African people can be perceived as two equally important 'characters' in directing Marlow's search and 'descent' to Colonel Kurtz:

> At night sometimes the roll of drums behind the curtain of trees would run up the river and remain sustained faintly, as if hovering in the air high over our heads, till the first break of day. Whether it meant war, peace, or prayer we could not tell. The dawns were heralded by the descent of a chill stillness; the wood-cutters slept, their fires burned low; the snapping of a twig would make you start. We were wanderers on prehistoric earth, on an earth that wore the aspect of an unknown planet. We could have fancied ourselves the first of men taking possession of an accursed inheritance, to be subdued at the cost of profound anguish and of excessive toil.
>
> (Conrad 1989: 68)

It seems that in the European perspective on Africa, its people only get shape, meaning and personality against the physical background of the landscape. Images of an African landscape are a European's necessary, but also paradoxical and ambivalent context for describing, and relating to, Africa(ns).

The paradox and ambivalence in these images is of a complex nature, however. For instance, much talking about Africa(ns) has a tendency to include the whole of Africa and to be taking into account all Africans, while some cultures and landscapes in Africa were and are more favoured by Europeans than others. With regard to southern Africa, it was especially the Zulu that captured the European imagination. As Draper and Maré (2003) drawing on Kiernan make clear, '(n)o other African people caught the Western imagination more powerfully than the Zulu. The land of the Zulu was extensively exoticized in literature and historical writing – and later in film and television' (Draper and Maré 2003: 553).

For that matter, it is not just Africans being stereotyped against a backdrop of certain landscapes. Europeans in Africa too are portrayed in a stereotypical way against the same type of background as the masculine descriptions of fearless European explorers and hunters indicate. The white male Europeans in the popular literature of Wilbur Smith and the like acquire their masculinity solely in the context of their surviving the African landscape, in which descriptions of fierce wildlife, harsh climate and pristine landscapes play an important role. As Martin Hall says about the Lost City, a (mainly European and American) tourist resort where European images of Africa(ns) appear to have been the root inspiration for the designers, architects, and programmers (see go2africa: 2002):

> The world has only one role for Africa – as a destiny for other people's expectations, and as the home of 'dark forces'. Rider Haggard [writer of amongst others 'King Solomon's Mines'], Wilbur Smith and Sol Kerzner [entrepreneur exploiting Sun City as part of his Sun International chain] have all seen this point – and have become wealthy.
>
> (Draper and Maré 2003: 557)

All of these authors and entrepreneurs built their fortunes on the ambivalent European imagery of African landscapes.

Europeans developed a similar paradoxical image of the people living in these landscapes: the Africans. On the one hand the African was considered an authentic 'noble savage', and on the other hand a violent and promiscuous barbarian. These images are in fact two sides of the same coin: an image of what Erlmann has described as 'spectatorial lust' (Erlmann 1999: 109–13). Elsewhere this is described as 'a carnival act consciously designed to play up their abnormalities – i.e. their radical deviation from European norms of dress and behavior' (Lindfors 1999: 77). According to Corbey (1989), these 'abnormalities' can be categorised into four main themes, dominating the European image of the African Other during the colonial era: violence, sexuality, eating habits and dress codes. Corbey lavishly illustrates these categories with examples of French postcards from that period. On these cards Africans were portrayed as violent warriors, capable of the worst acts of barbaric violence towards their enemies and towards Europeans.

African male Others were often seen as a sexual danger towards white women (see Shephard 2003). 'Judgements about the sexual behaviour of the people colonized by Europe played a core part in cultural othering and were central to the representation of the "horror" to use Kurtz's term, that was seen' (Channock 2000: 20). African women were usually considered to be sexually willing and the Europeans have always stereotyped the eating habits of Africans as cannibalistic. The well-known stereotype of Africans dancing around the cooking-pot while preparing Europeans for their dinner is a classic example. Finally, the Africans, especially in the context of Victorian prudence, were stereotyped as underdressed. The fact that women were walking around bare-breasted was 'proof' of their state of primitiveness and strongly reinforced the image of sexual willingness of African women. At the same time, by way of contrast, this reinforced the Western idea of its own superior civilisation (Corbey 1989).

Indeed, the aesthetic aspect of the African Other came to the fore in European image of African women. During the colonial era the number of male Europeans travelling to and through Africa was far larger than that of European women. In depicting the African Other, African women played an important role in the sense that they were 'measured' and contrasted to European aesthetic standards of women (pp. 39–40). One of the important representations of this European aesthetic imagery of African women was photography, and more specifically what Corbey (1989) has labelled the 'colonial nude'. This type of photography was especially dominant in representations on postcards sent to Europe from French and Belgian colonies in black Africa. German, English and Portuguese postcards of colonial nudes are not as common. This can possibly be explained by differences in Christian confessionals between these countries and the differences in the ascribed role of sexuality and eroticism (p. 23), but that is beyond the scope of this chapter.

European standards of (ethnocentric) aesthetics, both with regard to landscape and African people, played an important role in representations through photographs, literature and art on Africa and Africans in Europe. Most representations depict the paradoxical attitude of fear and attraction. This combination was often purposefully used in visual and literary representations to keep the audience and/or readership interested: fascinating but at a safe distance from reality. In this fashion a complete 'Otherness-industry' (Schipper 1995: 9) emerged in Europe in the second half of the eighteenth century to represent Africa and the Africans to a European audience. The Otherness-industry manifested itself in Europe in various forms, including photographs/postcards, museums, travel-literature and world exhibitions, that is, all elements of the 'industrialization of image making' (Crawshaw and Urry 1997: 182). All manifestations were primarily directed at Europeans who did not have the opportunity or did not dare to travel to Africa themselves, but who nevertheless wanted to experience 'the authentic Africa' through the representations and comments of people who had been there.

'Being there' was an accepted 'proof', never mind how strange the stories sometimes sounded in the European context. On the contrary, the more exotic and deviating from European standards the representation, the more possible and plausible it seemed. (For just how far 'out of their minds' many of these European explorers were and how that affected their representations of Africa(ns) and for what (un)reasons, see Fabian 2000.) The higher the contrast that was suggested, the more it seemed to verify European superiority. In that sense the African Other became 'extremalized', taking on a hyper-reality for the sake of contrast to European standards. Romantic aesthetics were the major undercurrent of these forms of representation, which put 'authentic' Africa and Africans on a European stage.

Africa(ns) on stage in Europe

During the second half of the nineteenth century the stage was set for showing Africa and Africans to Western civilisation through 'World Exhibitions'. The first exhibition was held in London in 1851 at Crystal Palace, specifically built for the occasion. It was followed, among others, by Paris (1855 and 1900), Vienna (1873), Philadelphia (1876), Amsterdam (1883), Antwerp (1885) and so on (see Corbey 1993). These exhibitions coincided to a large extent with the peak of postcard-sending in Europe (Crawshaw and Urry 1997: 185). The postcard was introduced in Austria in 1869, followed in 1870 by the *Korrespondenz-Karte* in Germany. Within a couple of years all other European countries followed suit. Two months after the introduction of the postcard in Germany, already two million had been sent. In 1889, more than fifty million postcards were sent in France alone (Corbey 1989: 16). The postcard proved a perfect stage for representing Africa and Africans to a Western audience. As the camera 'cannot lie', the idea was that it represented the African Other in its most pure and most authentic form. 'Photographic (. . .) apparatuses permit retention of the memory of the encounter, of "being there", and of iconically affirming the "visibility" by which the Others (re-)present themselves' (Tomaselli 2001: 176).

As mentioned above, however, there were striking uniform themes dominating the postcards sent from Africa to the European audience: violence, sexuality, eating habits and dress codes. This was not Africa and Africans in all its social and natural complexity: this was Africa(ns) reduced and reshaped to dominating Western expectations and modes of controlling and appropriating the African space and the Africans themselves. By way of categorising their behaviour into these four themes, the Otherness was neutralised and a European meaning was 'glued' onto it. African reality was made to fit to a European discourse on Africa, by way of the violent framing through the photographic image.

One of the wider known examples is that of an African woman who appeared on stage in world exhibitions and who featured in many visual representations of Europeans, the Khoisan Saartjie Baartman (as told in Corbey 1989 and Strother 1999). Far from being 'only' an example from times gone by, she is still part and parcel of a struggle between European imagery of and domination over Africa and African self-determination. Saartjie Baartman was a Hottentot woman who was first presented to the European audience in 1810 in London, from which she toured the English provinces and Paris as a sensational curiosity and representative of Africa and African women. She died in 1815, at 25 years of age, because of an infection. She suffered from steatopygia, enlargement of the behind, which was shown naked when 'on stage' and which stimulated many (male) sexual fantasies among the European spectators: 'If they paid a little extra, visitors could touch her buttocks' (Lamprecht and Baartman 2003: n.p.). She came to be known as the Hottentot Venus. At that time, the Hottentot/Bushmen were considered by anthropologists to be the race closest to primate monkeys, together with Australian Aborigines. In other words, on a hierarchical scale from the superior Western Caucasian race to primate monkeys, the Bushmen featured close to the lowest hier-archical level. This combination of steatopygia and status as the lowest ranking race secured for the exploiters of Saartjie Baartman a tremendous interest from the European audience. In France, the most respected anato-mist at the time, George Cuvier, took a keen interest in her bodily compo-sition. When alive she never allowed the famous scientist an analysis of her private parts, in which he was interested most, because of what has popularly become known as the 'Hottentot apron', or 'hypertrophy of the labia minora'. Only after she died was he able to analyse them in detail. 'The final product of Cuvier's artists' examination depicts a landscape in which the photographic realism of the woman's three-quarter portrait fuses seamlessly with a *beautiful landscape* full of such 'typical' Venus figures' (Strother 1999: 34–5, italics added).

But this was not yet the end of Saartjie Baartman, either in terms of artistic display or in terms of academic attention. A plaster cast of Saartjie Baartman appeared on display at the *Musée de l'Homme* in Paris until 1982. From the 1980s onward she began to draw new scientific attention in the literature on nineteenth-century exhibitions of people, and especially through the work of Stephen Jay Gould and Sander Gilman on the construc-tions of sexuality in science and medicine (Strother 1999). In 1993, a dispute arose after the Orsay museum in France had tried to redisplay the plaster cast of Saartjie Baartman. But South African organisations began demanding that her remains be returned to South Africa instead. Their demands have recently become more urgent since the remains of a man, only known as 'El Negro', have been returned to Botswana, where he was given a 'decent burial'. In an attempt to get back what was originally theirs but taken during the years of 'colonial plunder', full attention has now turned to Saartjie Baartman.

The French proposed that her remains be granted to a South African museum on an extended loan, rather than be repatriated for a dignified burial. But such a move would have to be endorsed by the French parliament, which regards the museum's contents as national treasures. . . . The French have denied that Baartman had been treated badly and kept in France against her will in the 19th century, and rejected claims that body parts were being kept in jars. [Cecil le Fleur], who represented the Khoi-Khoi Indigenous First Nations of South Africa organization [pleaded:] 'She had to display her posterior and genitalia in order to amuse callous, inhumane, insensitive crowds and white audiences as one of their peculiar finds in Africa'.

(Anon. 2000)

After many negotiations between France and South Africa, the remains of Saartjie were finally returned to Africa, after 192 years, on 3 May 2002.

Saartjie Baartman can be seen as an extreme example of the imagination of Europe about Africa. Saartjie Baartman was an African tourist attraction in Europe, both representing its people and even its landscape in an aesthetic context of comparison and contrast. Africa and Africans were, in the representation of Saartjie Baartman, confined to the few general categories in which the European tourist was able to capture the African continent (i.e. violence, sexuality, eating habits and dress codes), and (partly) through science this violent moulding of Africa to European categorisation was 'proved' and legitimised.

Primitive art

Another strong example of Africa(ns) on stage in Europe is the Western perception of so-called 'primitive art' in ethnographic museums (Price 1989). '(A)rt, specifically so-called primitive art, seems best to reflect in contemporary consciousness the idea of Africa' (Mudimbe 1994: 55), as '(c)ultures . . . have always branded themselves through art and ritual' (Channock 2000: 27). Here I do not mean to write about a 'primitive' surge in modern art, which inspired European artists such as Gauguin and Picasso. Rather, for the purposes of this chapter, the referral to primitive art is defined in the following: 'We are dealing with the arts of people whose mechanical knowledge is scanty – the people without wheels' (Price 1989: 2) or 'primitive art is produced by people who have not developed any form of writing' (p. 2). In the words of Price (p. 5), 'it deals, in short, with some of our most basic and unquestioned cultural assumptions – our "received wisdom" – about the boundaries between "us" and "them"'.

In European perceptions of primitive art the same paradoxical attitude can be witnessed as was earlier described for landscape and Africans: on the one hand admiration for the 'natural', 'instinctive' and 'basic' approach of art by the primitive, that is, African, artist. The Rousseauian ideal of the noble savage can be discerned again here. On the other side, primitive art has always been associated in the West with the more dark side of the Other. This again is the Otherness stigma that can immediately be taken from Conrad's *Heart of Darkness*:

> Doubtless, a bloodier eccentricity was never conceived by human madness: crimes continually committed in broad sunlight for the sole satisfaction of god-ridden nightmares, of terrifying ghosts! The priests' cannibalistic repasts, the ceremonies with cadavers and rivers of blood – more than one historical happening evokes the stunning debaucheries described by the illustrious Marquis the Sade.
>
> (Price 1989: 39)

Price (1989) describes the noble savage under the heading of 'The Universality Principle'. She argues that in a world which is becoming smaller and smaller all the time because of technological innovation, there is a strong tendency of Westerners to think of the world as one global family, enjoying equality. This Universality Principle is most strongly promoted and used for marketing by companies such as Coca-Cola and Benetton. They present a happy world with people of all shades of colour smiling to each other and into the camera. In this idea, art is supposed to play a major role in unifying the people of the world. Art, just like music and sport, brings together people from all corners of the world: '(a)rt in all its forms has been historically the most enduring language for the mingling of souls in common enjoyment . . .' (p. 29). According to Price, the universality of primitive art in this respect is that it pours out of the very universal depths of our human souls, psychological drives and exist-ence. 'Primitive artists are imagined to express their feelings free from the intrusive overlay of learned behaviour and conscious constraints that mould the work of the Civilized artist' (p. 32). This is how primitive art unites people from around the world, and this is how Africans and Europeans can meet as equals.

However, 'from the privileged perspectives of white Europeans and Americans, the mingling of races strongly implies an act of tolerance, kind-ness and charity' (p. 25) (probably similar to the 'kindness' and tolerance shown by Westerners in educational and pedagogical programmes towards children, models of intercultural communication and development aid). Similar to the European image of the pristine landscapes of Africa, prim-itive artists were considered as 'purified bearers of the human unconscious, as survivors of our lost innocence' (p. 33). In other words, a pristine psychological landscape of Eden is depicted, fitting into and matching the African natural landscape.

The other side of the coin of the noble savage in art, according to Price (1989), however, is the 'pagan cannibal', and 'the imagery used to convey primitive artists' otherness employs a standard rhetoric of fear, darkness, pagan spirits, and eroticism' – a remarkable overlap with the categories discerned and described by Corbey (1989). Price (1989: 45) also stresses explicitly that 'sexuality is clearly another important aspect of the image of Primitives as "the night side of man"'. This aspect is considered one of the major selling points of primitive art as a primitive arts dealer once told Price: '(o)bjects that are strongly sexed sell well' (p. 47).

Besides sexuality, there are two other issues which Price (1989) raises in relation to primitive art: anonymity and timelessness of primitive art. Both issues bear a broader meaning for the European image of Africa with which the tourist industry has to deal. In art, the name of the artist is usually partly responsible for the perception of the public and the price of a piece. People go to the Van Gogh Museum in Amsterdam because they consider Van Gogh a great artist. His famous sunflower painting was sold at an incredibly high price a few years ago, largely because it was an 'authentic Van Gogh'. The artist's name, authorship and the periodisation of the piece of art can literally be worth millions. This holds for painters and also for other artists such as sculptors and composers. In the case of primitive art, on the other hand, we can see two things: first, the name of the artist is hardly ever known. Secondly, the exact periodisation of the piece of art is not considered important, the latter because it is considered to be in line with 'age-old' traditions, in which individuals should be merely considered as torchbearers of collective traits, not as individuals with their own specific creativity. Personal inventiveness is not considered as an option for 'primitive artists'.

Price (1989: 61) comes to the cynical observation that 'once having determined that the arts of Africa . . . are produced by anonymous artists who are expressing communal concerns through instinctual processes based in the lower parts of the brain, it is but a quick step to the assertion that they are characterized by an absence of historical change', in other words anonymity and timelessness. It is important to mention these two issues here, especially because the earlier case of Saartjie Baartman seems to suggest otherwise – that there is individuality in the European image of Africa and Africans. But Saartjie Baartman only became known as a *representative* of all African women, not as an individual with unique traits. Saartjie Baartman was the strictly coincidental choice out of 'strictly similar' options of other African women. She did not become known because she was Saartjie Baartman, but because a European specifically chose her to represent African women and gave her a European name.

The same process of Europeans showing the European tourist the 'authentic and timeless Africa' can be clearly distilled from the case of the Bushmen in southern Africa. This case starts during the colonial era and is still in full process in present-day tourism developments in southern Africa. It is therefore a fitting final case in making plausible that European imagery with regard to Africa(ns) is not something to safely confine to the old days of colonial blindness and prejudice, but is still an important and potent aspect of European perceptions of present-day Africa(ns).

The Bushmen of southern Africa: from exhibition on stage to exhibition on location

The Bushmen of southern Africa are possibly the most exploited example of the ambivalent European image of Africa(ns), Perceived on the one hand as a people who are closer to primate apes than to Europeans, and hailed on the other hand as the 'noble savage' living in complete harmony with the natural African environment. As a result, the distance between myth and reality was stretched to its utter limits: '(f)ar from being "beautiful people living in a primeval paradise", they are in reality the most victimized and brutalized people in the bloody history that is southern Africa' (Gordon 2000: 10). (See Silvester and Gewald (2003) for an account of the brutalities afflicted on Bushman (and other 'natives' in South West Africa, now Namibia) during German colonial rule in a recent reprinted local account and report of it, the so-called *Blue Book*, originally written in 1918 and based on 50 African witness.)

During the early years of colonisation of South West Africa by the Germans around the turn of the twentieth century, Bushmen were literally hunted, in the same fashion as vermin. Bushmen at that stage were seen as a 'plague' (Gordon 2000: 57). In a sworn statement, a farmer admitted that he 'accompanied the German police and troops when they used to hunt Bushmen' (p. 77). It was partly through the influence and created imagery of anthropology that the Bushmen became a romantic curiosity, to be conserved in an especially created reserve for reasons of their language and physical constitution. Chilvers, in 1928 (in Gordon 2000: 148), 'reckoned the Bushmen as one of the Seven Wonders of southern Africa'. They had to be conserved just like the Taj Mahal in India. Two major influences can be discerned in creating, polishing and legitimising the current Bushmen myth. The first is Donald Bain, a big-game hunter and the second is anthropological science.

In 1936, there was an Empire Exhibition in Johannesburg in South Africa. While in Europe the ethnological exhibits lost their attraction to the broader public in the 1920s, in South Africa they continued to attract (white) people until the 1950s. In 1952 there was the Van Riebeeck Festival celebrating 300 years of European settlement in South Africa and the display of Bushmen was one of the highlights of the festival (Gordon 1999). The imagery of Europeans emigrating to and living in South Africa can be perceived as constituting a bridge between European imagery during the colonial era and the new interest for African culture that we experience in tourism today. The image of the pristine African landscape and the people who fit into that landscape is clearly articulated in the text accompanying the exhibits:

> These people were one time sole owners of Africa – the only living beings who could speak, kindle a fire and fashion implements. Their

signature in the shape of rock carvings and paintings is writ large over the face of Africa. All that today remains to them of their mighty heritage is a small portion of the Kalahari desert and the primeval forest to the West of the Albert Nyanza.

(Gordon 1999: 281)

One of the reasons Bain exhibited Bushmen to a larger audience was to persuade the government to grant them a 'Bushmen reserve'. Almost natural allies for Bain in his struggle to create a special Bushmen reserve were scientists. According to Gordon (1995: 320) 'Scientists had developed a vested interest in maintaining and indeed elaborating on the acceptable conventional wisdom concerning the Bushmen . . . (s)cientists formed the core of the various public support committees'. The first to join Bain were scientists from the Witwatersrand University, starting with Raymond Dart (Gordon 1999: 273, after Gordon 1995). This proved to be a start of a line of mainly anthropologists who constructed an image of the Bushmen which still holds today in (cultural) tourism and the popular perception of the Bushmen in Europe. In a photo-book entitled *Children of the Kalahari*, Alice Mertens for instance, describes a Bushman-girl she pictures in the following words, relating her to the landscape she (seems to) belong in, in *optima forma*:

Unkra, a girl with a skin as smooth and coppery brown as a ripe berry, and a smile as bright as the early morning over the Kalahari veld. Unkra, whose name sounds like the cracking of a nut shell, was born in the year of the good rains when game and veldkos [bushfood] were plentiful and all the Bushmen were happy and content.

(Mertens 1966: Preface)

The pictures that come with the rest of the narrative seem solely selected to illustrate this harmonious relation between people and landscape. The fitting last words of the book are: '(a)nother fine day has passed and Unkra and Xua watch the glow of the sun getting softer and the dark shadows getting longer over the golden grass of the Kalahari' (Mertens 1966: n.p.).

The history of San studies from the 1950s onward shows an impressive list of scientists who devoted their time and energy to the study of the San in a vein similar to Mertens, that is, 'intellectuals romanticizing traditional ways of life' (Dalby 2002: 131). Scientists doing research on the Bushmen included Lee and De Vore (1968), Kuper (1970), Marshall (1976) and Shostak (1981). The Marshall family is probably the most well-known family creating a romantic image of the bushman. 'The Marshalls' work did much to establish an international image of the Bushmen as an ancient people, unspoiled, living in harmony with their environment' (Jones 2001: 211). However, it was Laurens van der Post (1958) who has probably been the most influential in bringing the Bushmen onto the global stage through his best-selling books on the Bushmen, of which the most well known is probably *The Lost World of the Kalahari*. 'In the romantic perspective which Laurens did so much to echo and establish, the Bushmen men were hunters and the women gathered tubers and fruit and leaves from the Kalahari bush' (Jones 2001: 212). In his review of Jones' biography, Attwell (2003: 308) comes to the conclusion that Laurens' 'life expressed its aesthetic alternative as an existential, lived reality', that is, the European colonial gaze of Africa in the flesh. As with most Europeans, van der Post was also selective in his idolising of noble Africans. He too divided them into 'good natives' and 'bad natives', to borrow Neumann's (2000) terminology.

> Good natives are those having a 'traditional' livelihood sustained by 'indigenous knowledge'. They are perceived to be closer to nature and thus consistent with the environmental managers' design for parks [and aesthetics of landscape]. Bad natives are those who are in some sense 'modern', and thus removed from the nature, their modified lifestyles and greed for consumer goods representing a particular threat to the natural treasures enclosed [and thus not fitting in the imaginary African landscape].
>
> (Broch-Due 2000: 29)

As Birkett (1997: n.p.) concludes in her article years ahead of Jones' biography, 'Van der Post found his ideal Africans, not in nationalists and freedom fighters, but in the far less threatening Bushmen'.

One thing all of these studies and books have in common is that they make extensive use of descriptions of landscape in picturing the Bushmen. It seems that it is only against this landscape that the life of the Bushmen can be understood. The word landscape itself is not often used, because it was not part of the early anthropological discourse, but it was translated into concepts such as 'environment' (Marshall), 'life in the bush' (Shostak) or 'ecology' (Lee and De Vore). The history and tradition of the San studies is also of interest to note as it indicates how much of the anthropological discipline was still working within a format developed during the colonial era. It depicted an evolutionary continuum with stages from primitiveness to civilisation, and from non-human primates to European whites. In other words, it was Darwinism applied to the social world.

In the early 1960s, the anthropological world was excited by the new data pouring in from field studies of non-human primates and from the Leakeys' discoveries of ancient living floors associated with *fossil man*. The ethnographic study of a contemporary hunter-gatherer group seemed to be the next logical step (Lee and De Vore 1976: 10). No wonder the Bushmen were sometimes referred to as 'living fossils' (as said by General Smuts, as acting Prime Minister) (Gordon 1995: 32).

Based on the above explications, it is possible to conclude with Gordon (2000: 250) that 'there is little difference between the current and past scientific and popular images of Bushmen [and that] the overwhelming textbook image is that they are *different* from us in terms of physiognomy, social organization, values, and personality'. In other words, Bushmen can be perceived as the 'Ultimate Other', a mirror through which Europeans can contrast and measure their own achievements.

Moving the stage from Europe to Africa in the European quest for authenticity

In our European search for authenticity and the authentic experience we have replaced the stage for the African Other from Europe's World Exhibitions, journals, scientific ethnographies, *National Geographics*, television documentaries, and so on, to Africa itself. While the stage may have been replaced, however, the format and the script of the imagery have been left intact. Africans still have to '*perform* Europeans' prior ideas of them' (Landau 2002: 19). Europeans want to see the Africans and the African landscape in the same way as they are taught to see them in their formative years of image moulding during the colonial period. Therefore, Europeans long for immaculate African landscapes with picturesque thatched roofs dotted and merging into it, and expect to hear the drums the minute they arrive in Africa, with Africans rhythmically dancing to their ongoing cadenza. That is Africa. That is the Otherness (i.e. 'them') Europeans (i.e. 'us') want to experience in Africa and for which they are prepared to pay money. This is the imagery or staged authenticity to which the tour operators have to relate in their brochures in order to persuade clients/tourists to book a holiday with them. This is the imagery of African culture which cultural tourism must reflect in its programmes. In the words of Dalby (2002), this might be called 'landscape consumption' (p. 166).

The appropriate symbol for this land consumption, which makes the consumption comfortably possible, is the 4X4:

> The domination of nature [and landscape for that matter] within a specifically colonial sensibility, is what many of the vehicle advertisements are all about. . . . In Britain, Landrover's Freelander SUV [Sports Utility Vehicle] is released into the wild by African park wardens.
>
> (Dalby 2002: 168–9)

The 4X4 is presented as the ultimate transporter to reach the stages set in far off and exotic landscapes. One of these 'authentic stages' is the Kgalagadi TransFrontier Conservation Area (TFCA), located in South Africa and Botswana. A 4X4 is the preferred mode of transport for reaching the various destinations in this TFCA. The Peace Parks Foundation (PPF) sponsors the development of the tourism potential in this TFCA. One of their projects was Project 26/2, to initiate a San Cultural Centre. The project consisted of a workshop, a so-called 'Mobile Cultural Tourism Workshop'. It was meant 'to examine various existing cultural tourism businesses as well as develop new ideas that would allow communities *to showcase their culture* in a financially and culturally sustainable way' (Peace Parks Foundation n.d.).

The stage has changed from Europe to Africa itself, but the accompanying text, discourse and associated imagery remain firmly the same. As Anderson and Grove had argued already in 1987 (in Beinhart 1987: 4), it 'remains true that Europeans and their ideas exert an undiminished, even increasing influence over the African environment today'. In the same book Beinart (p. 17) observes that 'the Western world draws on old-established strands in ideas about Africa: a contradictory set of attitudes that would at once see Africa developed and wild'.

Further to the above, however, it has been argued in this chapter that from a European cultural perspective, African landscapes and people cannot be separated from each other with the landscape taking priority over the people. The people belong to the landscape; the landscape constitutes the people. It is the African environment and landscape into which Africans have to fit seamlessly to satisfy the European spectator and attract them to the stage in Africa as overseas tourists willing to pay vast amounts of money, i.e. foreign exchange, for their images coming to life. Africans have to fit in, in order to be considered 'good natives'. What this means for rethinking development issues from a European perspective, i.e. how well modernity fits African landscapes, is relevant to consider in the context of Africa being the number one recipient of the Western aid-business, but is beyond the scope of this work. Nevertheless, in their search for (staged) authenticity Europeans 'go into Africa in the hope of discovering a strand that might root them more firmly into the earth and the cosmos (. . .) [and which offers them] cosy refuges from the chilling winds of modernity' (Draper and Maré 2003: 564, 559). Drawing on the colonial period, their perspectives of African reality are (often) bent in order to fit their images and sense of 'staged authenticity' of the African landscape and its people.

Acknowledgements

Thanks to Malcolm Draper for comments and suggestions on an earlier draft of this chapter.

References

Anderson, D. and Grove, R. (1987) 'The scramble for Eden: past, present and future in African conservation', in D. Anderson and R. Grove (eds) *Conservation in Africa: People Policies and Practice*, Cambridge: Cambridge University Press, pp. 1–12.

Anon. (2000) 'El Negro is home – will Saartjie be next?' *The Star* 6, October: n.p.

Anon. (2003) 'Preserving indigenous knowledge and culture', *The Natal Witness*, 16 July: n.p.

Attwell, D. (2003) 'Telling stories: whites seeking home in Africa, review of Jones' biography of Laurens van der Post', *Journal of Southern African Studies* 29, 1: 307–9.

Beinart, W. (1987) 'Introduction', in D. Anderson and R. Grove (eds) *Conservation in Africa: People, Policies and Practices*, Cambridge: Cambridge University Press, pp. 15–20.

Birkett, D. (1997) 'Close look reveals Sir Laurens van der Posture', *Weekly Mail and Guardian*, 19 December 1997.

Broch-Due, V. (2000) 'Producing nature and poverty in Africa: an introduction', in V. Broch-Due and R. A. Schroeder (eds) *Producing Nature and Poverty in Africa*, Stockholm: Nordiska Afrikainstitutet, 9–52.

Bunn, D. (1999) 'An unnatural state: Tourism, water, and wildlife photography in the early Kruger National Park', paper presented at the conference *African Environments: Past and Present*, 5–8 July, St Anthony's College, Oxford University.

Carruthers, J. (1995) *The Kruger National Park: A Social and Political History*, Pietermaritzburg: University of Natal Press.

Channock, M. (2000) '"Culture" and human rights: orientalising occidentalising and authenticity', in M. Mamdani (ed.) *Beyond Rights Talk and Culture Talk: Comparative Essays on the Politics of Rights and Culture*, Cape Town: David Phillips Publishers, pp. 16–36.

Conrad, J. (1989 (1902)) *Heart of Darkness*, London: Penguin Books.

Corbey, R. (1989) *Wildheid an beschaving. De Europese verbeelding van Afrika.* (translation: *Savagery and Civilization. The European Image of Africa*), Baarn: Ambo.

Corbey, R. (1993) 'Ethnographic showcases, 1870–1930', *Cultural Anthropology* 8, 3: 338–69.

Cosgrove, D. E. (1984) *Social Formation and Symbolic Landscape*, London, Sydney: Croom Helm.

Crawshaw, C. and Urry, J. (1997) 'Tourism and the photographic eye', in C. Rojek and J. Urry (eds) *Tourism Cultures: Transformations of Travel and Theory*, London, New York: Routledge, pp. 176–95.

Dalby, S. (2002) *Environmental Security*, London, Minneapolis, MN: University of Minnesota Press.

Doy, G. (1998) *Women and Visual Culture in 19th Century France, 1800–1852*, Leicester: Leicester University Press.

Draper, M. (2003) 'Going native? Trout and settling identity in a rainbow nation', *Historia* 48, 1: 55–94.

Draper, M. and Maré, G. (2003) 'Going in: The garden of England's gaming zookeeper and Zululand', *Journal of Southern African Studies* 29,2: 551–69.

Erlmann, V. (1999) '"Spectatorial lust": The African choir in England, 1891–1893', in B. Lindfors (ed.) *Africans on Stage: Studies in Ethnological Show Business*, Bloomington, IL and Indianapolis, IN: Indiana University Press, Cape Town: David Phillips Publishers, pp. 107–34.

Fabian, J. (2000) *Out of Their Minds. Reason and Madness in the Exploration of Central Africa*, Berkeley, CA: University of California Press.

go2africa (2002) *Sun City: Africa's Kingdom of Pleasure*, www.sun-city-south-africa.com/palace.asp [accessed 8 August 2003].

Gordon, R. J. (1995) 'Saving the last South African Bushman: a spectacular failure', *Critical Arts. A Journal of Cultural Studies* 9, 2: 32.

Gordon, R. (1999) '"Bain's bushmen": Scenes at the Empire Exhibition, 1936', in B. Lindfors (ed.) *Africans on Stage. Studies in Ethnological Show Business*, Bloomington, IL and Indianapolis, IN: Indiana University Press, Cape Town: David Phillips Publishers, pp. 266–90.

Gordon, R. (2000) *The Bushman Myth. The Making of a Namibian Underclass*, 2nd edn, Boulder, CO: Westview Press.

Gordon, R. (2002) '"Captured on film": Bushmen and the claptrap of performative primitives', in P. S. Landau and D. D. Kaspin (eds) *Images and Empires. Visuality in Colonial and Postcolonial Africa*, Berkeley, CA: University of California Press, pp. 212–32.

Grove, R. (1987) 'Early themes in African conservation: the Cape in the nineteenth century', in D. Anderson and R. Grove (eds) *Conservation in Africa. People, Policies and Practices*. Cambridge: Cambridge University Press.

Grove, R. (1997) 'Scotland in South Africa: John Croumbie Brown and the roots of settler environmentalism', in T. Griffiths and L. Robin (eds) *Ecology and Empire: Environmental History of Settler Societies*, Pietermaritzburg: University of Natal Press, Keele University Press.

Jahoda, G. (1999) *Images of Savages: Ancient roots of modern prejudice in Western culture*, London, New York: Routledge.

Jones, J. D. F. (2001) *Storyteller: The Many Lives of Laurens van der Post*, London: John Murray.

Kirkaldy, E. N. (2003) 'The darkness within the light: Berlin missionaries and the landscape of Vendaland c. 1870–1900', *Historia*, 48, 1: 169–202.

Kuper, A. (1970) *Kalahari Village Politics. An African Democracy*, Cambridge: Cambridge University Press.

Lamprecht, A. and Baartman, S. (2003) *Amagugu: South African Heritage Training and Technology Program*, www.saculturalheritage.org (accessed 15 August 2003).

Landau, P. S. (2002) 'Introduction', in P. S. Landau and D. D. Kaspin (eds) (2002) *Images and Empires: Visuality in Colonial and Postcolonial Africa*, Berkeley, London: University of California Press, pp. 1–40.

Landau, P. S. and Kaspin, D. D. (eds) (2002) *Images and Empires: Visuality in Colonial and Postcolonial Africa*. Berkeley, CA and London: University of California Press.

Lee, R. B. and De Vore, I. (eds) (1968) *Man the Hunter*, Chicago: Aldine.

Lee, R. B. and De Vore, I. (eds) (1976) *Kalahari Hunter-gatherers. Studies of the !Kung San and their Neighbors*, Cambridge, MA: Harvard University Press.

Lemaire, T. (2002) *Mit open zinnen. Natuur, landschap, aarde (translation: With Open Senses: Nature, Landscape, Earth)*, Baarn: Ambo.

Lindfors, B. (ed.) (1999) *Africans on Stage. Studies in Ethnological Show Business*, Bloomington, IL and Indianapolis, IN: Indiana University Press; Cape Town: David Phillips Publishing.

Lübbren, N. and Crouch, D. (eds) (2003) *Visual Culture and Tourism*, Oxford: Berg Publishers.

Marshall, L. J. (1976) *The !Kung of Nyae Nyae*, Cambridge, MA: Harvard University Press.

Maxwell, A. (2000) *Colonial Photography and Exhibitions: Representation of the 'Native' People and the Making of European Identities*, Leicester: Leicester University Press.

Mertens, A. (1966) *Children of the Kalahari*, London: Collins.

Mudimbe, V. Y. (1994) *The Idea of Africa*, Bloomington, IL and Indianapolis, IN: Indiana University Press, London: James Curry.

Neumann, R. P. (1998) *Imposing Wilderness: Struggles over Livelihood and Nature Preservation in Africa*, London: University of California Press, pp. 220–42.

Neumann R.P. (2000) 'Primitive ideas: Protected Area Buffer Zones and the politics of land in Africa', in V. Broch-Due and R. A. Schroeder (eds) *Producing Nature and Poverty in Africa*, Stockholm: Nordiska Afrikainstitutet.

Peace Parks Foundation (n.d.) 'Peace Parks Foundation projects supported and executed', unpublished.

Pickering, M. (2001) *Stereotyping. The Politics of Representation*, Basingstoke: Palgrave Macmillan.

Price, S. (1989) *Primitive Art in Civilized Places*, Chicago: University of Chicago Press.

Ranger, T. O. (1999) *Voices from the Rocks: Nature, Culture and History in the Matopos Hills of Zimbabwe*, Oxford: James Currey.

Rojek, C. (1995) *Decentring Leisure: Rethinking Leisure Theory*, London, New Delhi: Sage Publications.

Rojek, C. (1997) 'Indexing, dragging and the social construction of tourist sights', in Rojek, C. and Urry, J. (ed.) *Touring Cultures: Transformations of Travel and Theory*, London, New York: Routledge, pp. 52–74.

Schipper, M. (1995) *De boomstam en de krokodil. Kwesties van ras cultuur en wetenschap* (translation: *The tree-trunk and the Crocodile. Matters of Race, Culture and Science*), Amsterdam: Van Gennep.

Shephard, B. (2003) *Kitty and the Prince*, Johannesburg, Cape Town: Jonathan Ball Publishers.

Shostak, M. (1981) *Nisa: The Life and Words of a !Kung Woman*, London: Allen Lane.

Silvester, J. and Gewald, J. B. (2003) *Words Cannot be Found: German colonial rule in Nambia: An Annotated Reprint of the 1918 Blue Book*, Leiden: Brill Publishers.

Strother, Z. S. (1999) 'Display of the body Hottentot' in B. Lindfors (ed.) *Africans on Stage. Studies in Ethnological Show Business*, Bloomington, IL and Indianapolis, IN: Indiana University Press, Cape Town: David Phillips Publishing, pp. 1–61.

Tomaselli, K. G. (2001) 'The semiotics of anthropological authenticity: the film apparatus and cultural accommodation', *Visual Anthropology* 14: 173–83.

van der Post, L. (1958) *The Lost World of the Kalahari*, London: Hogarth Press.

Wilmsen, E. N. (1995) 'First people? Images and imaginations in South African iconography', *Critical Arts: A Journal of Cultural Studies* 9, 2.

Wolmer, W. (forthcoming) 'Wilderness gained, wilderness lost: Wildlife management and land occupation in Zimbabwe's southeast lowveld', *Journal of Historical Geography*.

6 Commodifying heritage

Post-apartheid monuments and cultural tourism in South Africa

Sabine Marschall

> The interlocking dimensions of time and space make the journey a potent metaphor that symbolizes the simultaneous discovery of self and the Other. It is precisely this capacity for mirroring the inner and the outer dimensions that makes possible the 'inward voyage', whereby a movement through geographical space is transformed into an analogue for the process of introspection.
>
> (Galani-Moutafi 2000: 205)

Introduction

Monuments, memorials and heritage sites are a way of taking a metaphorical journey through time and space. They represent an authorised, institutionalised interpretation of history; a public acknowledgement of loss, suffering and achievements; a recognition and validation of cultural identity. Heritage is a key mechanism in defining community, ethnic or national identity and re-inscribing the postcolonial landscape. Constructing identity often involves introspection, an 'inward journey', in Galini-Moutafi's words, a look into the past, an inspection and discovery of the Self, in order to determine who we are and where we come from.

South Africa is currently fascinated – if not obsessed – with the identification, celebration, re-evaluation and, not least, commodification of 'heritage'. While certainly a wider, international trend, this strong preoccupation has manifested itself in South Africa only with the advent of the post-apartheid dispensation, which is often equated (although not uncontested) with the postcolonial era. In this country – caught in a fundamental process of socio-political transformation, where the previously marginalised are now (politically) empowered – the key motivations for heritage are nation-building; the desire to tell the 'other' side of the story; the need to express new values and contribute to cultural empowerment. Yet, for the ordinary person 'on the ground', struggling for survival in the face of poverty and lack of opportunities, it is an entirely different argument that matters: Monuments and heritage sites – so it is claimed – contribute to economic empowerment through the attraction of tourists.

In South Africa tourism is currently promoted as the panacea of all ills. There is rarely a post-apartheid monument that is not expected to become the catalyst of development, employment and poverty alleviation through the attraction of tourists. The reality, however, often looks rather different. Considering the wide spectrum of monuments and heritage 'products' that range from simple memorials commemorating victims of township 'massacres' to the highly commercialised venture of a theme park nature, as currently under construction in Durban, this chapter will take a critical look at the issue of heritage and cultural tourism in the postcolonial/post-apartheid South African context. It will investigate to what extent monuments and heritage sites indeed attract tourists and development; where the challenges lie in commodifying memories; and what problems may occur if the commodification of heritage becomes too successful. Perhaps most importantly, it intends to raise awareness of the negative impacts that an uncritical and uncontrolled embrace of tourism may have on local identities and the authenticity of their cultural heritage.

Fascination with heritage

South Africa's current preoccupation with heritage is reflected in multifarious ways: at the legislative level in the form of new laws and policies, most notably the passing of the South African Heritage Resources Act in 1999; in the re-arranging and 'transformation' of museums throughout the country; in the annual celebration of 'Heritage Day' (24 September) with an array of associated rituals, festivals and exhibitions; in the popular enthusiasm for renaming streets and cities; and – last but not least – in the erection of countless new monuments and memorials throughout the country.

This fascination with heritage, which may involve not only the preservation, but sometimes the actual re-creation of the past, is a broader, international phenomenon, which has manifested itself in Europe, the United States and other parts of the world, long before it became predominant in South Africa. One might say it is a condition that characterises our contemporary age – psychologically linked to a deeply felt cultural need, a sense of nostalgia, which Nora (1989) explains through the loss of *mileux de mémoires,* 'environments of memory' that were once an intricate part of society.

There is another, more pragmatic, rationale behind the prevailing fascination with heritage, which certainly plays an important role in the current post-apartheid South African context. It is the desire to create a new national identity through a process of selective remembering, thereby simultaneously legitimating the present socio-political order. Heritage sites, and most notably monuments, become the visual manifestations of this official, public interpretation of the past. Yet, considering the reality of scarce resources in a country where a large percentage of the population lives below the poverty line, the considerable expenditure for such commemorative endeavours (e.g. R350 million for Freedom Park, to be built outside Pretoria as the country's foremost post-apartheid monument) is not easily justifiable. It is in this context that the tourism argument comes to fruition. Virtually all new monuments and memorials will – if official statements and press reports are to be trusted – attract hordes of cultural tourists, thereby contributing to infrastructure development, job creation and income generation for previously disadvantaged communities (for example, see Bishop (1998) about the Ncome monument near Dundee; Edwards (2000) about Sharpeville; Mkhize (2001) about monuments in Durban; and Moya (1997) about Soweto).

The motivational double-bind that drives the absorption with heritage – political and economic expediency – was poignantly expressed by South African Tourism Minister, Valli Moosa, who links heritage, tourism and economic development with the wider project of nation-building and identity in a post-apartheid society. In an article entitled 'Building a nation through our heritage', Moosa refers to the three World Heritage sites in South Africa declared in 1998:

> They are symbols or icons of what we as a nation can feel justifiably proud about in the world. We must take them and boldly start to project ourselves as a nation internationally whether through promoting investment or marketing tourism. . . . We have to start working on a consensus of how we see and want to build our nation. . . . The manner in which we do this cannot be separated from the process of nation building. We cannot say that our campaign to market SA to potential British tourists can be separated from nation building.
>
> (Ibid. 1998)

Monuments and heritage sites are meant to be visited; they are designed for the visitor, including the foreign visitor, the traveller or the tourist. Monuments and heritage sites are thus a vehicle for nation building, for constructing a new identity, and presenting this identity to the outside. The foreign (usually European) tourist as Other looks in and helps define the South African Self.

Heritage, postcolonialism and tourism

Writing in 1995, Thornton (1996) discusses Appiah's (1991) well-known article about the postmodern and the postcolonial and investigates how these terms might be applied to South Africa after the first general elections. Arguing that apartheid was 'a form of rampant modernism' (Thornton 1996: 136), which can furthermore be labelled 'postcolonial', Thornton concludes that post-apartheid South Africa can be called post-modern, but not postcolonial. Without engaging with the question of whether or not South Africa is postmodern, it is argued here – in contrast to Thornton – that post-apartheid South Africa is postcolonial, while the apartheid era was not.

Strictly speaking, South Africa has, of course, been a postcolonial nation ever since the political system of British colonialism formally ceased in 1910. However, it is now widely accepted that postcolonialism has more to do with power constructs than linear time. For example, consider Duncan (2002), who stressed that the term 'postcolonial' does not necessarily automatically apply to any country after attainment of political independence from a former coloniser. 'When *The Wretched of the Earth* was published in 1967, to use only one glaring African example, South Africa still lay under the colonial construct of apartheid regardless of legal independence from the former empire' (Duncan 2002: 325).

In the South African case, postcolonialism should more accurately be defined from the perspective of the vast population majority, whose domination and marginalisation was not ended, but – if anything – increased by the transition from British colonialism to South African union and then further to the apartheid state. The project of labelling and classification, and the moving of theoretical discourses from one historical or geographical context to another, invariably involves misfits and stretching of definitions (Noyes 2000: 52). Yet, for the purposes of this chapter, the advent of the post-apartheid era in South Africa, which formally ended the socio-political domination of the black population majority, is equated with the postcolonial era. The term 'coloniser', then, refers to members of the white minority regime, while the 'postcolonial agents' are those previously marginalised who have become empowered to 'speak'. Post-apartheid monuments and heritage sites are, in that sense, postcolonial monuments.

One of the key aims of the postcolonial project in any context is the (re)discovery of previously marginalised or denied history, the celebration and validation of previously disparaged values and other such strategies that contribute to the 'decolonisation of the mind'. Post-apartheid monuments and heritage sites are certainly postcolonial in that sense. They are also intended to break the monologue of the official historical record, to counter the biased accounts of the coloniser, to tell the 'other' side of the story. Heritage is thus a means of the postcolonial agent to assert a new (decolonised) identity.

If this contributes to clarifying the relationship between heritage and postcolonialism, what is the connection between postcolonialism and tourism, especially cultural and heritage tourism? A bulk of literature has emerged on tourism and postmodernism (e.g. Ritzer and Liska 1997; Urry 1990), yet scarce attention has been paid to the relation between tourism and postcolonialism – a fact acknowledged by the very initiative of this book. Colonialism, with its civilising mission and economic agenda, resulted in homogenising people with different cultures throughout a region or even – to some extent – worldwide. Postcolonialism, precisely through the concept of heritage, aims to counter the Western mould by fostering and, indeed, sometimes actually recreating, the traditions and cultural identities that were previously invalidated or suppressed.

Being financially under pressure, South Africa, like other postcolonial nations, attempts to find economically profitable, commercial applications of its celebration of cultural diversity through cultural and heritage tourism 'products' (heritage sites, craft, dances, festivals, cultural villages, etc.). While European colonialism, through radical exploitation of resources, is generally considered the root cause of their current economic weakness and outright poverty, the postcolonial developing countries now seek to exploit the (frequently European) foreign tourist in return. Alternatively, one may say, the postcolonial country exploits its own cultures in the pursuit of making money, perpetuating in the present – in a different form – the process started by the coloniser of the past.

Tourism tends to be enthusiastically promoted by government agencies (certainly in South Africa), but researchers are much less certain about whether it should be considered a blessing or rather a curse for the societies of developing countries. As a global economic force, tourism is still strongly dominated by and dependent on the West, which largely controls the international tourism industry, prompting some scholars to consider tourism a form of imperialism or neocolonialism (e.g. Nash 1977). On the other hand, it cannot be denied that tourism often has a positive effect on conservation – both of nature and of cultural heritage. The real question is not whether or not the postcolonial nation should promote tourism, but rather *what kind* of tourism should be encouraged and how this should be done.

What kind of heritage attracts tourists?

Cultural heritage is considered a key component of South Africa's tourism 'product', but what kind of heritage attracts tourists? While Valli Moosa (1998) advocates focusing on the positive aspects of the country's chequered past in projecting an image to the international community, others suggest that foreign visitors should also 'see the dark side of our history' (Jayiya 1998). This contention emerged in the context of Thabo Mbeki's inauguration as president in 1998, when the colonial and apartheid era statues on the grounds of the Union Buildings in Pretoria were covered in black cloth for the occasion.

When the tourism argument is mobilised to justify new monuments and heritage sites, the implied assumption is that tourists are attracted primarily or exclusively to the heritage of 'the people', meaning the African or non-white population. This is clear in the following quotation from Thembinkosi Ngcobo, eThekwini's Executive Director of Parks, Recreation and Culture, who, in advocating new monuments, explains:

Durban still looks like a colonial city and not much has changed since colonial rule. Tourists do not come here to see a mini London but an African city and how its people live. We need to Africanize the city. We are not saying all colonial statues and monuments should be removed. . . . They must be complemented by statues and monuments of other people who also played a role in the shaping of Durban.

(Ngcobo quoted in Mkize 2001)

Yet cultural tourists tend to favour a holistic, politically balanced, contextualised representation that allows them to understand the complex realities that have shaped a country's history and its people. There is evidence that monuments and heritage sites associated with 'the dark side of our history' are by no means unattractive to tourists. With the former stigma removed, the Voortrekker Monument outside Pretoria or the Afrikaner Taalmonument at Paarl, for instance, have become popular sites for both domestic and foreign tourists. An article in the *Daily News*, contextualising the monument issue, remarked the following about the Afrikaans Language Monument at Paarl:

Do these sweeping pillars of crushed granite and concrete celebrate the language alone, or rather the nationalism that was built up around it? Ample tribute is paid to the combined influences of Europe and Africa in its symbolism, but when the monument was first conceived, the intention clearly was to link the Afrikaans language with an expression of nationalism. . . . Unexpectedly perhaps, the monument is taking on a life of its own which goes way beyond its original conception. Today it is actively visited, mainly by tourists of Dutch and German extraction keen to explore their cultural links with South Africa, now that they are no longer constrained by the stigma of apartheid.

(Anon. 1999)

At site visits, local staff at the Taalmonument and at the Voortrekker Monument at Pretoria confirmed the popularity and attractiveness of these sites to tourists. In fact, a statistical analysis might reveal that more foreign tourists visit the Taalmonument (celebrating the Afrikaans language) at any one time than the post-apartheid Ncome monument near Dundee or the Sharpeville memorial. The Afrikaner 'homeland' of Orania in the Northern Cape, where a Verwoerd statue – dismantled in Bloemfontein – has proudly been re-erected, was recently featured in *Sawubona* (2002), the South African Airways (SAA) inflight magazine, implicitly marketing the place as a foreign tourist attraction.

Do new, post-apartheid monuments indeed attract tourists? Is it true that the erection of new monuments and heritage sites leads to local development and economic prosperity for previously disadvantaged communities? These are critical questions – rarely investigated – which warrant closer examination.

Visual appearance

When the author recently presented a paper on post-apartheid monuments at an international conference, the first questions, by a delegate from France, were: 'Why are they so boring? Why do they imitate European conventions?' The majority of post-apartheid monuments are indeed aesthetically, completely unoriginal and uncreative, imitating Western, and, in fact colonial formal conventions. Scarce resources might be a blessing in disguise, because if it was not for this restraint, every city would by now probably be dotted with a host of new busts and bronze statues on pedestals, featuring a variety of 'liberation heroes'.

The mimicry of Western models of commemoration may certainly strike one as a contradiction in the current post-apartheid context, when South Africa is searching for a new, African-based identity and purporting to strive towards the 'African Renaissance'. Tourism brochures and travel literature lead the overseas visitor to expect a unique African experience, full of exoticism and 'difference'. In fact, the attraction of the Taalmonument may in part lie in its suggestion of the 'otherness' of the country. As Spiegel (1994: 191) observes: 'Paradoxically, many such tourists visit the country for a taste of the very magical mystery and traditionality of Africa – the "otherness" – that the *taalmonument* suggests the continent and its people possess.' Ngcobo's comment quoted earlier suggested that new monuments are needed to 'Africanise' the city of Durban. 'When foreign tourists come to South Africa', says Ngcobo (personal communication 2002), 'they don't want to be reminded of having conquered us; they want to see an independent African identity asserting itself'. Interestingly, Ngcobo's statements drew a lot of attention and local press coverage at the time, one of which even compared the initiative with the Cultural Revolution in China:

> In a campaign almost reminiscent of the '50s Cultural Revolution initiated by Chairman Mao Dze Dong in Communist China, the eThekwini Municipality's Parks, Recreation and Culture Department is embarking on a drive that will see Durban becoming a truly African city.
>
> (Baloyi 2001)

A contradiction on the one hand, the common practice of imitating Western models can be explained by drawing on postcolonial theory. Much of postcolonial discourse has focused on literature and has described the literary text as a site of control, which, under colonialism, has effectively fixed the 'native' under the sign of the 'Other'. Monuments, operating through visual and textual language, are another form of such texts. Colonial and apartheid era monuments – by and large dedicated to the great master-narrative of 'civilisation' and progress brought by the European – have contributed to disseminating specific discourses around black Africans and other non-white populations in Southern Africa. As mentioned earlier, monuments of the post-apartheid period intend to deconstruct, subvert or invert the coloniser's discourses. The post-apartheid monument that mimics or imitates Western models can thus be interpreted as a strategy of the postcolonial agent to appropriate the (visual) language of the coloniser in order to 'write back' (Ashcroft *et al.* 1989), to respond to and 'de-scribe' the discourses of the coloniser (Marschall 2003a).

While the lack of creativity and the Eurocentricity of the visual language of post-apartheid commemorative structures and sculptures can thus be explained, it is likely to diminish considerably their attractiveness to foreign tourists. Cultural tourists, in particular, tend to be widely travelled. Their frame of reference with respect to monuments may be informed by examples of conceptually or aesthetically sophisticated – sometimes very resource-intensive – commemorative projects encountered during their varied journeys. Being surrounded – in their home country – by countless cenotaphs, memorial steles and statues on pedestals accumulated over centuries or indeed millennia, it can be surmised that new South African commemorative efforts are unlikely to grab any tourist's imagination.

Focus on content

Defenders of the new monuments would argue that it is not the visual appearance that is meant to attract tourists, but the content of what they represent. In other words, monuments and memorials function as landmarks, alerting visitors to an important event that has occurred there, thus enabling them to experience the spirit of the place, to trigger their imagination and to vacuously partake in a truth that has shaped history. Without necessarily admiring its aesthetic, the monument or statue nevertheless allows visitors to pay their respects to the heroic leader it commemorates.

Indeed, it appears that tourists do visit new monuments and heritage sites – at least to some extent. There are a number of heritage site developments that have become very successful as tourist attractions, most notably Robben Island. Another great 'seller' among South African heritage 'products' is the site of the June 1976 Riots in Soweto, commemorated by the Hector Pieterson Memorial and adjacent museum – now a standard item of every Soweto township tour. Although it is not quite evident in which way local residents benefit from the tourist flow, the memorial has certainly brought along some development and general upgrading of the area and it can be anticipated that this process will continue in future. The link between monument and development is even more pronounced at the proposed Walter Sisulu Square in Kliptown. Here a monument is planned to commemorate the open square where a popular mass meeting was held in 1955, leading to the adoption of the Freedom Charter. The monument is linked to a substantial urban renewal project for the entire surrounding area, intended to provide roads, homes and public facilities. The master plan – winning entry (by Johannesburg-based firm StudioMAS) of an open competition – indeed looks impressive.

Incomplete monuments

This long-term development project, driven by the Gauteng provincial government, is envisaged to be accomplished in several phases, the first of which will be the completion of the monument in time for the fiftieth anniversary of the historic event in 2005. Although provincial authorities have earmarked a substantial sum (over R400 million) for the purpose, a closer look reveals that this amount does not cover all aspects of the project. For instance, separate funds must still be raised for the restoration of historic buildings identified by the South African Heritage Resources Agency (SAHRA) as worthy of conservation (Khumalo, personal conversation 2003). Chances are that the hype and flurry of activity running up to the completion of the monument, is likely to be followed by a (perhaps temporary) phase of inertia and lack of motivation following its unveiling – exacerbated by shortage of funds, bureaucratic procrastination and other inhibiting factors.

This suspicion is justified on the basis of previous experience with post-apartheid monuments. Ravi Jhupsee (personal conversation 2003), architect of the Resistance Park monument in Durban, recalls the passionate and feverish bustle shortly before the structure was unveiled by former president, Nelson Mandela, in May 2002. This included a remarkable eagerness, among institutions and private sponsors alike, to come forward with donations, ensuring that their name would remain affiliated with this important (and politically correct) initiative. Yet, after the official unveiling, the situation took a marked turn. The monument still stands incomplete today; few sponsors can be motivated to alleviate the crucial lack of funds and, in fact, the first signs of deterioration are already visible due to the poor workmanship of the rushed job. The monument has not (yet) been included in the major tourist routes or the repertoire of 'must see' attractions reflected in tourist brochures.

At the Samora Machel memorial in Mbuzini near the Mozambique border, the South African government's failure to deliver on its agreement for the fencing in of the memorial structure has almost led to a diplomatic incident. Mozambican government officials urged that the memorial needed to be protected from vandalism by local residents and complained that 'each time we come to the monument, we find cattle relieving themselves' (Anon. 2000b) – not exactly an attractive prospect for foreign tourists, one might think!

At the Ncome monument near Dundee, which commemorates the Zulu victims of the famous battle of Blood River, a bridge was to be built, linking the Afrikaner Blood River monument with the new Zulu monument (Coan 1998; Pienaar 1998). This was not only meant to facilitate convenient access for tourists, but to fulfil a symbolic function as an expression of the post-apartheid 'rainbow nation' spirit and reconciliation among former enemies. The fact that the bridge has still not been built thus takes on a more profound symbolic significance, apart from having obvious practical implications. Needless to say, tourist numbers at Ncome are hardly impressive and the scores of begging children descending upon the visitor suggest that the development and poverty alleviation objectives associated with this project have been struggling to succeed. In fact, one may question the impact of this development, and the 'rich' tourists it draws, on the moral fibre of this rural community – given that a group of begging children might be able to make more money from tourists than their mothers can through producing crafts for sale. However, this is not meant to suggest that the Ncome project has completely failed in its development and poverty alleviation objectives. A proper socio-economic study would be needed to find out, precisely and objectively, to what extent the local community has benefited. Even the presence of begging children is not necessarily a reliable indicator of the level of poverty in an area.

Challenges in commodifying heritage

Despite the attractiveness of certain icons of Afrikaner nationalism mentioned earlier, it can – without doubt – be assumed that the heritage of African and other non-white communities is potentially of much greater interest to tourists, certainly foreign tourists. Yet, there are a number of challenges in turning the heritage of 'the people' into tourist attractions. As Hynes (1999: 59) rightly asks: 'Can one commodify and sell memories? And who is empowered to remember – and for whom?' If monuments are visual manifestations of memories, then whose memories do they represent? As much as, for instance, Sharpeville is hailed as a promising 'political tourism' site, the question of 'who owns the memory of Sharpeville' (Edwards 2000), has been a contentious one even before the advent of the post-apartheid period (e.g. Anon. 2000a; Frankel 2001; Ngidi *et al.* 2002). Fuelled by the recent monument project, issues around ownership and representation clearly have the potential to divide communities.

The question of representation is a very sensitive one in South Africa today and there is acute awareness that the history and culture of previously underprivileged groups in society have always – in the past – been presented by the privileged. The advent of the new dispensation has brought with it a strong move to change this pattern, to strive for a larger degree of self-representation and to emphasise community participation. To some extent, this has led to a complete inversion of the former practice, whereby the racial or ethnic affiliation of the curator or project leader is regarded as an automatic qualification to accurately and authentically represent the culture, history and interests of his/her community.

As mentioned earlier, the current obsession with 'completing the record' and inclusiveness is aimed at adding an African perspective to the previous Eurocentric record. Yet, this inclusiveness has led to new (or rather continued) absences and exclusions, notably with respect to women (Marschall

2003b). The present, strong under-representation of women's experiences and female heroes in post-apartheid monuments, memorials and statues is a result, not necessarily of ill-will, but of the frequent male-dominant composition of heritage committees and the dominance of their male agendas.

One of the key challenges in commodifying the heritage of 'the people' lies in the frequent lack of material substance of that heritage. The former National Monuments Council (NMC) had focused its attention on the conservation of 'monuments' in the sense of solid, built structures – many of which became popular tourist attractions without much preparation for the 'tourist gaze'. The fact that African communities have traditionally produced few 'monuments' in this sense was addressed through the dissolution of the NMC and the establishment of the South African Heritage Resources Agency (SAHRA) in 1999. The shift from the term 'monuments' to 'heritage' was widely interpreted as a progressive move as it opened up the field to include a broad range of objects and sites – not necessarily containing any built structures. Yet, tourism thrives on the visual experience and on the tangible. How do we successfully commodify intangible aspects of heritage and sites where 'there is nothing to see'?

The erection of a monument might be a suitable strategy to fill the gap. The monument becomes a marker in space, alerting the tourist or passer-by (as mentioned earlier) to an important event that has occurred here. Some monuments are even conceptualised as viewing platforms, from which the visitor can gaze upon the precise site where the event took place. This applies, for instance, to the Langa Massacre Memorial at Uitenhage, commemorating the fatal shooting incident that took place here at the local township in 1985. The memorial thus not only commemorates the dead, but it brings the event to life by encouraging visitors to imagine, to 'see' in their minds what happened. As Kirshenblatt-Gimblett (1998: 166) has accurately observed, 'Heritage and tourism show what cannot be seen – except through them'.

As in the case of the Langa Massacre Memorial, most of the significant events that are being commemorated – and potentially commodified for the tourist – are associated with townships or remote rural areas. This fact poses further challenges – notably in terms of access and security. 'Security could prove to be the major test of the council's resolve to make Soweto a major tourist destination after seven Swiss tourists were recently robbed while visiting Regina Mundi', reported *The Star* (Moya 1997). Likewise, *Business Day* reported that bustling tourism trade is going on at the Shaka memorial in Dukuza ('You can also buy genuine Zulu assegais' (Lee 1999)), but further ventures into the Zulu interior to other sites associated with Shaka were not exactly recommended, unless with a guide. At Resistance Park in Durban, the visitor frequently encounters vagrants loitering around the park or sleeping between the pillars of the monument. In all these cases (and there are many more), the resources are simply not available to ensure the safety and security of visitors. As tour operators are usually not prepared to take risks, many of these heritage sites may be left off the tourist itinerary – as attractive as they might be.

Nelson Mandela as tourist attraction

One way of solving such problems associated with the tourist's trip to a heritage site, is to bring the heritage to the tourist. This may involve erecting monuments that are primarily addressed at tourists, set up in places designed for tourists. The bronze statue of Nelson Mandela at Hammanskraal is an example of this type. The small town of Hammanskraal near Pretoria has recently been furbished with a new 'centre' at the town's fringe. Here a series of solid craft stalls have been built along a new street – wide enough for tourist coaches – and around the traffic circle, where a bronze statue of Nelson Mandela forms the focal point. This arrangement allows the tourist to shop for curios, watch the makers of the craft items at work, and take a picture of South Africa's foremost iconic personality all at the same time, without having to worry about the inconveniences and security risks of actual city life.

April 2004 also saw the unveiling of the latest Mandela statue in Sandon Square, now renamed Nelson Mandela Square. Indeed, there has been a strong trend towards cashing in on the international popularity and iconic stature of Nelson Mandela since the beginning of the post-apartheid era. Other proposals have included the so-called Freedom Monument. This emerged in 1995 and envisaged a giant bronze cast of Mandela's hand, breaking through prison bars. The 23-metre (some sources say 33-metre) high sculpture, was to be privately funded by businessmen Solly and Abe Krog at a cost of R50 million and sculpted, ironically, by Danie de Jager, an artist closely associated with the commemorative endeavours of the apartheid regime.

The project drew an unprecedented amount of debate and criticism (see, for example, Anon. 1995; Anon. 1996; Greig 1996; Vanderhaeghen 1996; Dubow 1996). The concept is 'in the best tradition of fascist South African monumental kitsch', commented Robert Greig (1996), Arts Editor of the *Sunday Independent*. Similarly, Neville Dubow (1996) argued that the monument employs the language of the 'discredited rhetoric of totalitarian art':

> That monumental arm that is supposed to symbolize freedom, bursting through prison bars, is it waving or drowning? In its overblown, vein-bulging literalism, it is an echo of all that is bad in the discredited rhetoric of totalitarian art. It is blatantly the wrong image for the nation we are trying to build.
>
> (Ibid.)

Furthermore, he says:

> It is the language of the instant sell; the language of the theme park.
> It is precisely the language of the backers of the project, the brothers
> Krok, who talk in terms of the largest cast bronze sculpture in the
> world and the *Guinness Book of Records*, and how many tourists it
> will attract. Is this what we want?
>
> (Ibid.)

With respect to the monument's tourism potential, Marilyn Martin, director
of the South African National Gallery, added: 'Tourists would indeed flock
to see the monument, but to laugh at South Africans' naivety and philis-
tinism, not to share in their liberation through a work of art' (Martin,
quoted in Vanderhaeghen 1996).

The Mandela Hand project was eventually called off, but the idea has
obviously inspired the more recent proposal for an even more gigantic
statue of Mandela with his arm raised. Imitating the Statue of Liberty in
New York, it would exceed this model in height by almost 20 metres (van
Heerden 2001; van Niekerk 2001; Philp 2002). The monument is envisaged for the coastline at Port Elizabeth, allowing that city to solidify its
identity as gateway to the Eastern Cape – heartland of Nelson Mandela
and the forces of the liberation movement. Driven by a local businessman
(Kenny McDonald), the entire monument complex will cost about R1
billion and is intended to become South Africa's foremost tourist attraction, drawing a projected 5,000 tourists per day (Philp 2002). The statue
is meant to rotate and will be equipped with all the trappings of a successful,
commercial tourist enterprise according to Western standards, including a
restaurant and conference centre and a wax museum à la Madame Tussauds
in London. (The current status of this project is that R2 million have been
set aside to conduct a full feasibility study; if finally approved, the project
is due to be completed by 2006 (Philp 2002)).

Despite efforts to prevent Mandela from being turned into a commodity,
such initiatives and their strong association with tourism implicitly serve
to trivialise the man's role and personality and prepare the way for his
likeness to be turned into an item of kitsch. The trend towards trivialisa-
tion and commodification of national icons is replicated at regional and
community level, most notably with respect to the historic figures of King
Shaka Zulu and Mahatma Gandhi. Development proposals for the Durban
beachfront include a statue of Prophet Isaiah Shembe and a Ghandi monument. The latter, it was suggested, should be 'part of an historical park
where all the leaders in the struggle for peace and human dignity against
apartheid should be depicted in their original attire and background' (Pillay
2000).

Commodification of Zulu heritage

In the province of KwaZulu Natal (KZN), now branded 'The Kingdom of the Zulu' by KZN Tourism Authority, monuments celebrating Zulu heritage (e.g. the new 'Spirit of Makhosini' monument near Ulundi) and statues to Zulu kings (especially Shaka), quite obviously cater for the tourists' sense of exoticism, thereby reinforcing stereotypes created by the coloniser. The bronze statue of Shaka Zulu in front of the former KwaZulu Legislative Assembly at Ulundi is based on the well-known illustration by Nathaniel Isaacs, published in 1836, which has been proclaimed the only 'true' likeness of Shaka. However, Dan Wylie (2000) has meticulously analysed this image and most convincingly deconstructed its claim to authenticity, highlighting that virtually no secure facts exist about Shaka or what he looked like. This image, Wylie (p. 13) contends, is less about Shaka than it is about a 'reverse representation of who the *European* is, or thinks "he" is'.

Nevertheless, the stereotypical image is deeply anchored in the public imagination and largely taken for authentic by tourists and local communities, including Zulu speakers, alike. The tourism industry in South Africa (as much as elsewhere) has been thriving on highly stereotypical representations of local cultural identities, and it is not surprising that KZN Tourism Authority has chosen Isaac's iconic colonial image as part of its logo, contributing to its further dissemination nationally and internationally. This is not to imply that no statues should be erected to Shaka because we do not know exactly what he looked like. Rather, this is to advocate a higher level of critical awareness about the way stereotypes are created through the constant repetition of the same model. Such stereotypes can set standards for what people consider authentic – not only about Shaka Zulu, but perhaps about their own sense of identity as Zulus.

Preliminary sketches suggest that the image might once again become the model for a proposed Shaka statue in Durban (Kearney 2001). This statue was envisaged to be part of uShaka Island, the multi-million Rand theme park development opened in April 2004 at the Durban beach front. Featuring marine life and Zulu cultural heritage, the park is part and parcel of the city's larger urban development framework and the upgrading of the 'Golden Mile' and harbour front, which includes the nearby Sun Coast casino complex, new hotels, restaurants and upmarket shopping areas. The target audience of these developments is domestic and foreign tourists. uShaka Island is the ultimate in commercialisation and commodification of heritage, or what Ritzer and Liska (1997) call, its 'McDisneyization'. Here the heritage of King Shaka and the Zulu people is appropriated to become a kind of décor that lends a local flavour to the international-standard, commercial entertainment and shopping experience. Culture, says Jameson (1991), is the 'new logic' of capitalism. As mentioned earlier, heritage serves economic and political interests. In KZN, ruled by the

Inkatha Freedom Party (IFP), this exploitation of the economic potential of Zulu heritage, simultaneously and conveniently suits the party's traditional agenda of promoting Zulu nationalism.

It can be anticipated that the representation of Zulu heritage at uShaka Island will be less concerned with researched authenticity than with 'customer satisfaction' – the fulfilment of tourist fantasies and myths – stressing difference, otherness and exoticism. Most of these myths and fantasies have their origin in the colonial era, when the 'civilised' European tried to make sense of his encounter with the 'savage native'. As Wels (2000) has shown, many of the stereotypes and myths about the Other continue to be disseminated – in modified and updated form – by the promotional literature of the travel and tourism industry, associated visual products such as postcards and the popular media (e.g. coffee-table books and movies).

Conclusion

uShaka Island is an extreme case of commodification of heritage, which marks one end of the spectrum and is not (yet?) representative of South African heritage practices in general. As much as the current obsession with heritage might be criticised – especially within academia – it cannot be denied that the celebration and even the commodification of cultural heritage is widely perceived by ordinary people as empowering. The 2002 conference of the South African Historical Association (SAHA) with the theme 'Heritage Creation and Research: the Restructuring of Historical Studies in Southern Africa', for example, exposed a great deal of uncertainty about what exactly 'heritage' means, especially in the South African context. While many of the contributions by white academics manifested a highly critical or negative attitude, 'heritage' was clearly perceived as positive and empowering by some of the black African participants. For many previously oppressed communities, it represents a form of validation and acknowledgement – at long last – of their own culture and history. Monuments and heritage sites can be the means of self-expression or self-representation by the postcolonial agent, responding to the biased narratives of the coloniser. Ironically, the (politically/culturally) empowering aspect of heritage may even include the stereotypical images created by the coloniser, which are by no means being deconstructed, but – on the contrary – being embraced. The nostalgia for the 'uncontaminated Other' exists on the part of the tourist, as much as on the part of local communities, for whom the illusion of a pre-colonial past often elicits a sense of pride and self-esteem.

This highlights the complexities surrounding the role and discourses of the coloniser, and how to position them in the postcolonial context. Recent critiques have centred on post-colonialism's heavy reliance on difference and binary oppositions, as well as its polarised, dichotomous view of the relationship between coloniser and colonised. As Nuttall and Michael (2000: 10) observed: 'Post-colonial readings of culture have tended to focus on difference – but more complex studies of affinities and how they are made are now needed, particularly in South Africa.' Moreover, in relation to South Africa they specifically noted: 'Cultural theorizing in South Africa, with its emphasis on separation and segregation, has been based until recently on the following tendencies: the over-determination of the political, the inflation of resistance, and the fixation on race, or more particularly on racial supremacy and racial victimhood as a determinant of identity' (Nuttall and Michael 2000: 1–2). As Comaroff recently pointed out, the problem in black South Africa these days is that nobody any longer knows who or what the enemy is (Bhabha and Comaroff 2002: 45).

In summary, the celebration of heritage and its associated commemorative practices may be empowering in the current post-apartheid, postcolonial context. Furthermore, in a country of scarce resources, where tourism is one of the most important growth industries, it may be legitimate to cash in on the foreign attractiveness of local cultural heritage. However, upon closer examination, many current heritage projects may in reality have much lower value as tourist attractions and much lower potential for economic development than South African communities are being made to believe. In fact, one suspects that the tourism argument is sometimes mobilised to justify heritage projects that are rather politically expedient. At the opposite end of the scale, it may be equally problematic if heritage becomes too attractive for tourists. If pursued uncritically and uncontrolled, the appropriation of cultural heritage for commercial exploitation may cause more harm than good for local communities and South Africa as a nation.

References

Anon. (1995) 'Mandela sculpture', *The Mercury*, 2 April.

Anon. (1996) 'Bronze "idol" sheer waste', *Eastern Province Herald*, 12 April.

Anon. (1999) 'A monumental debate', *Daily News*, 4 January.

Anon. (2000a) ('Sapa') 'Government plans to build monument to honour Sharpeville dead', *Natal Witness*, 22 March.

Anon. (2000b) 'Mozambique hits at SA on crash memorial', *Pretoria News*, 23 October.

Appiah, K. A. (1991) 'Is the post- in postmodernism the post- in postcolonial?', *Critical Inquiry* 17, Winter: 336–57.

Ashcroft, B., Griffiths, G. and Tiffin, H. (1989) *The Empire Writes Back*, London: Routledge.

Baloyi, M. (2001) 'Durban shakes off its past', *The Independent on Saturday*, 17 November.

Bhabha, H. and Comaroff, J. (2002) 'Speaking of postcoloniality, in the continuous present: a conversation', in D. T. Goldberg and A. Quayson (eds) *Relocating Postcolonialism*, Oxford: Blackwell, pp. 15–46.

Bishop, C. (1998) 'Zulu monument still on track', *Natal Witness*, 18 December.

Coan, S. (1998) 'Blood River's many battles', *Natal Witness*, 16 December.

Dubow, N. (1996) 'Arms and the man', *Weekly Mail*, 12 April.

Duncan, D. (2002) 'A flexible foundation: constructing a postcolonial dialogue', in D. T. Goldberg and A. Quayson (eds) *Relocating Postcolonialism*, Oxford: Blackwell, pp. 320–33.

Edwards, K. (2000) 'Sharpeville will be one of SA's biggest political tourism sites', *Sunday Independent*, 25 June.

Frankel, P. (2001) *An Ordinary Atrocity: Sharpeville and its Massacre,* Johannesburg: Witwatersrand University Press.

Galani-Moutafi, V. (2000) 'The self and the other: traveler, ethnographer, tourist', *Annals of Tourism Research*. 27, 1: 203–24.

Greig, R. (1996) 'Kitsch is the Kroks' democratic right – but only if they keep it private', *Sunday Independent*, 7 April.

Hynes, V. (1999) 'Reconstructing identity: contemporary art in post-apartheid South Africa', *Artlink. Australian Contemporary Art Quarterly* 19, 3: 58–61.

Jameson, F. (1991) *Postmodernism, or the Cultural Logic of Late Capitalism*, London/Durham, NC: Verso/Duke University Press.

Jayiya, E. (1998) 'Blackout for apartheid statues', *The Star*, 14 June.

Kearney, L. (2001) 'uShaka Island project launched', *Natal Mercury*, 18 October.

Khumalo, V. (2003) South African Heritage Resources Agency (SAHRA), Johannesburg office, personal communication, January.

Kirshenblatt-Gimblett, B. (1998) *Destination Culture: Tourism, Museum and Heritage*, Berkeley, CA: University of California Press.

Lee, P. (1999) 'Shaka's burial place rediscovered', *Business Day*, 16 April.

Marschall, S. (2003a) 'Setting up a dialogue: monuments as a means of "writing back"', *Historia* 48, 1: 309–25.

Marschall, S. (2003b) 'Serving male agendas: two national women's monuments in South Africa', South African Association of Art Historians (SAAAH). Nineteenth annual conference, Stellenbosch, September.

Mkhize, T. (2001) 'Durban to take on "African city" look', *Sunday Times*, 27 May.

Moosa, M. V. (1998) 'Building a nation through our heritage', *Business Day*, 3 December.

Moya, F.-N. (1997) 'Council taps into tourism market', *The Star*, 6 March.

Nash, D. (1977) 'Tourism as a form of imperialism', in V. L. Smith (ed.) *Hosts and Guests. The Anthropology of Tourism*, Philadelphia, PA: University of Pennsylvania Press, pp. 33–47.

Ngcobo, T. (2002) eThekwini's Executive Director of Parks, Recreation and Culture, personal conversation, Durban, March.

Ngidi, T., Mntungwa, M. and SAPA (2002) 'PAC boycotts unveiling of Sharpeville Memorial', *The Mercury*, 22 March.

Nora, P. (1989) 'Between memory and history: Les *Lieux de Mémoire*'. *Representations* 26 (Spring): 7–25.

Noyes, J. (2000) 'The place of the human', in S. Nuttall and C. Michael (eds) *Senses of Culture: South African Culture Studies,* Cape Town: Oxford University Press Southern Africa, pp. 49–60.

Nuttall, S. and Michael, C. (eds) (2000) *Senses of Culture. South African Culture Studies,* Cape Town: Oxford University Press Southern Africa.

Philp, R. (2002) 'Giant Mandela statue planned', *Sunday Times*, 20 October.

Pienaar, H. (1998) 'Taking Blood River in the right vein', *The Star,* 12 December.

Pillay, D. G. (2000) 'Living monument to ancestors needed', *Daily News*, 7 February.

Ritzer, G. and Liska, A. (1997) '"McDisneyization" and "Post-tourism": complementary perspectives on contemporary tourism', in C. Rojek and J. Urry (eds) *Touring Cultures: Transformations of Travel and Theory*, London: Routledge, pp. 96–109.

Spiegel, A. D. (1994) 'Struggling with tradition in South Africa: the multivocality of images of the past', in G. C. Bond and A. Gilliam (eds) *Social Construction of the Past. Representation as Power,* London: Routledge.

Thornton, R. (1996) 'The potentials of boundaries in South Africa: steps towards a theory of the social edge', in R. Werbner and T. Ranger (eds) *Postcolonial Identities in Africa,* London: Zed Books, pp. 136–61.

Urry, J. (1990) *The Tourist Gaze: Leisure and Travel in Contemporary Societies,* London: Sage.

Vanderhaeghen, Y. (1996) 'Monumental questions', *Natal Witness*, 9 April.

van Heerden, D. (2001) 'Mandela-beeld kry gestalte', *Oos-Kaap Rapport,* 3 June.

van Niekerk, L. (2001) 'Oorsese hulp stroom in vir beeld', *Burger,* 10 July.

Wels, H. (2000) 'A critical reflection on cultural tourism in Africa: the power of European imagery', Paper delivered at the ATLAS Conference: Cultural tourism in Africa: strategies for the new millennium, Mombasa, Kenya, December.

Wylie, D. (2000) *Savage Delight. White myths of Shaka*, Pietermaritzburg: University of Natal Press.

7   Tourism and British colonial heritage in Malaysia and Singapore

Joan C. Henderson

Introduction

The interpretation of heritage and its presentation to visitors creates numerous dilemmas (Tunbridge and Ashworth 1996; Ashworth 2000; Graham *et al.* 2000), particularly when linked to a history of colonisation (Palmer 1994; Shaw and Jones 1997). Postcolonial societies have to confront recent experiences of occupation and external control with the potential for conflicts over the narratives to be communicated about the past to contemporary audiences, both domestic and international. The responses of the authorities to these challenges reflect wider agendas and the imperatives of nation building and constructing a sense of national identity which are frequently connected to hegemonic goals. There is also increasing recognition that heritage is a valuable tourism resource, and that heritage attractions, some with colonial associations, are highlighted by those responsible for destination marketing and development. Images and text are often employed in promotion to evoke nostalgia for an imperial age among markets from developed countries of the West, including erstwhile empire builders.

Understandings of heritage in general and its official articulation are thus affected by various dynamics, and several commentators have written about the dangers of its exploitation by tourism which is criticised as a vehicle for the perpetuation of colonial attitudes and structures (Britton 1982; Morgan and Pritchard 1998). The demands of tourists take precedence over those of the local community, and the tourism industry thus gives rise to inequalities and inequities which are based on assumptions of the superiority of Western culture and European-Atlantic dominance (Said 1994). However, it can be argued that tourism has the capacity to encourage a more balanced relationship between destination and generating countries, depending on how it is managed, so that neocolonial pressures may be resisted. Tourism also acts as a means of helping to bind individuals and groups together in the uncertain and insecure world of independence, reinforcing feelings of a common history and destiny through the identification and celebration of national symbols and stories (Palmer 1999).

The difficulties of dealing with heritage are apparent regarding any tangible colonial remains which often constitute a distinctive landscape, especially within an urban setting. Governments have to make decisions about this inheritance while establishing themselves and the fledgling state, and also in looking to the future. One problem is ensuring that buildings are accepted as belonging to the new nation and not the former coloniser. This problem is aggravated when buildings are perceived as an embodiment of oppression and the values of imperialism (King 1976; Southall 1971; Yeoh 1996). At the same time, some venues may have been the scene of critical events in the struggle for freedom (Simon 1984) and be deemed repositories of public memories which are worthy of protection, not least because of the favourable light such recollections cast on postcolonial authorities. Total obliteration is unlikely and often unrealistic, with the options of neglect or renaming and adaptation being more likely (Western 1985). Tourism provides an additional justification for conservation and reuse, with buildings being possible tourism assets which might yield economic returns.

This chapter explores postcolonial connections between heritage and tourism within the contexts of Malaysia and Singapore in south-east Asia whose histories as British colonies are briefly summarised. The discussion focuses on the built colonial heritage of Singapore's city centre and Georgetown and Kuching, the respective capitals of the Malaysian states of Penang and Sarawak, which all contain striking examples of architecture from the period of British rule. The ways in which this legacy is utilised and presented as a visitor attraction and its other roles are examined, with reference being made to the motives which underlie official approaches to conservation. Policies are shown to be exposed to influences in the broader political, social and economic arenas which determine and may distort the versions of colonial history which are formally sanctioned. It should be noted that it is not the intention within this chapter to debate definitions of postcolonialism (see Williams and Chrisman 1994). Rather, the term is used here to denote the status of comparatively newly independent states (Jacobs 1996).

The colonial legacy in Malaysia and Singapore

Trading opportunities drew the British government to the Malay region during the late eighteenth century when the East India Company was searching for Eastern ports in which to conduct business. Penang Island and Province Wellesley, an area of land on the mainland opposite, were ceded to the Company in 1786 and Stamford Raffles claimed Singapore as a trading post in 1819. Britain later acquired Malacca from the Dutch and these territories were united as the Straits Settlements, which was designated a Crown Colony in 1867 (Turnbull 1989). Elsewhere on the Malay Peninsula, the British operated a dual system of administration.

Resident generals sought to coordinate policy in the federated states which had their capital in Kuala Lumpur, and the unfederated states were governed indirectly by a series of British advisers to Malay rulers (Baker 1999). Sarawak on the island of Borneo in Eastern Malaysia was originally under the control of the Sultan of Brunei who rewarded the English adventurer, James Brooke, for his assistance in crushing a local rebellion by appointing him Rajah and Governor in 1841. The unique dynasty of the 'White Rajahs' (Tarling 1971; Payne 1986) was to last until 1946 when Sarawak too became a British Crown Colony (Porritt 1997).

The Second World War and Japanese invasion were crucial stages in the region's history, confirming disillusionment with the British regime and strengthening indigenous opposition. The Malayan Union, incorporating Penang, was formed in 1946 and Singapore became a separate Crown Colony after an interlude of British Military Administration Rule. The Federation of Malaya was founded in 1948 with a view to self-government. The years 1948 to 1960 were marred by the euphemistically entitled 'emergency' of Communist insurgency, and Malaya gained independence in 1957. Singapore was granted internal self-government in 1959 and, alongside Sarawak, joined the Federation of Malaysia on its establishment in 1963, but left in 1965 and was declared a fully independent republic (Chew and Lee 1991). The countries have since seen significant change, with Singapore now among the most prosperous in Asia and Malaysia also having undergone rapid advancement.

The British were thus a powerful presence in Malaysia and Singapore for over 100 years, shaping the physical landscape and erecting numerous buildings (Hayes Hoyt 1991, 1993; Jayapal 1992). Development intensified at the end of the nineteenth century in Peninsular Malaysia and Singapore, a consequence of the booming tin and rubber industries, and wood was replaced by more solid building materials. Communications were extended and an infrastructure of facilities and services installed as the population increased. Administrative and commercial offices, banks, railway stations, hospitals, private homes, clubs and schools were all constructed as well as residential accommodation. Urbanisation continued into the twentieth century, leading to the creation of public and private spaces which were reminiscent of parts of Britain and had little relation to the tropical surroundings. Designs varied from mock Tudor to Neo-Gothic (Gurnstein 1985; Edwards and Keys 1988; Vlatseas 1990), but the predominant style was derived from English Palladian Georgian architecture which was subsequently introduced into India where certain modifications were made. This has been categorised as Anglo-Indian and was 'meant to tell a moral tale' as 'monuments of rulers, and a way of life impervious to the riotous and excessive East' (Beamish and Ferguson 1985: 24), messages to be communicated throughout the British Empire.

The grandest and more intimidating physical edifices could be seen as an assertion of the primacy and power of the ruling class, with preservation thus unlikely to be awarded the highest priority in the postcolonial era. Despite reservations about their merits and future, however, practical necessity initially demanded the utilisation of some buildings in Malaysia and Singapore. They were later acknowledged as a source of valuable historical and cultural insights, being aesthetically pleasing (Liu 1996) and having an especial appeal to tourists. Official positions on colonial heritage sites have thus evolved over time, against the background of shifting opinions about heritage in general. The stances on conservation in the two countries are outlined in the next section.

Conserving colonial heritage

According to Powell (1994: 16), 'the developing world has often equated heritage – especially colonial heritage – with backwardness and considers things of the past as old baggage which should be discarded to achieve modern statehood'. This view appears to have prevailed initially in Malaysia and Singapore. Officials in Singapore contended that retaining old properties and districts was a waste of space and meant that land, already in short supply, was unable to maximise its commercial potential. Conservation could stand in the way of progress, thereby damaging the economy and national interest (Dale 1999).

Such attitudes were gradually revised as a result of the growing awareness of heritage's contribution to nation building and an apparent appreciation of its intrinsic worth. Singapore's racial mix of a predominantly Chinese population with substantial Malay and Indian minorities has given urgency to the task of realising an all-embracing identity which transcends ethnic allegiance, thereby averting conflict (Hill and Lian 1995). Feelings of a shared history, including the journey to freedom from colonial masters and subsequent advancement, are ties which hold disparate groups together and give them a stake in the country. The satisfaction of material needs has also led to greater attention being devoted to more abstract matters, one concern being the quality of urban life. These considerations underpin the government's present programme of restoration which aims to:

> add variety to our streetscapes and modulate the scale of our urban fabric, creating the visual contrast and excitement within the city while protecting the important reminders and representations of our past. In addition, it adds to the distinctive character and identity of our city, giving it a sense of history and memory of the place.
>
> (Urban Redevelopment Authority (Singapore) 2003)

The postcolonial government in Malaysia also faced numerous difficulties, one of the most intractable being to secure 'national unity within a heterogeneous plural society' (Watson 1996: 297) in which a Malay majority coexisted with Chinese and Indian communities. There were also strong regional loyalties and sometimes greater identification with traditional sultans than Malaysian nationhood. Religion was another obstacle to an all-encompassing nationalism for, although Islam was the official religion, almost half the population were not Muslim. Despite these complexities of identity, the Malays were privileged and government nurtured a 'Malay-centric national culture' (Cartier 1997: 577), which has obscured the contribution of minorities to Malaysian history. At the same time, forging social cohesion and a unifying identity has been pursued in order to guard against disintegration (Crouch 2001), and the common burden of colonialism has been harnessed on occasion to the cause of unity.

The hegemonic implications of heritage and its conservation must not be overlooked, as a stable society inculcated with officially approved accounts which are flattering to the ruling elite is less likely to challenge prevailing ideologies. Singapore's People's Action Party (PAP) has held office since independence and consistently positions itself as the architect of Singapore's successes as well as defender of the values on which the maintenance of prosperity depends (Rodan 1996). Its readings of history and their transmission through various media assist in reinforcing selected interpretations and undermining dissent. The prevailing force in Malaysian politics is the broad based Barisan National coalition led by the United Malays National Organisation (UMNO) party, headed by Mahatir Mohamad who was Prime Minister from 1981 to 2003. Mahatir has spearheaded the drive to modernisation and full development status which has transformed aspects of Malaysian life (Shamsul 1999). Nation building is thus fused with attempts at social and political control as the ruling party asserts that its own future and that of the modern state are inextricably linked, with stability and security demanding loyalty and electoral support.

The heightened priority allocated to conservation in Singapore (Smith 1999) is revealed in the emphasis given to it by the Urban Redevelopment Authority (URA), the agency in charge of planning, which has incorporated the topic into the republic's Master Plan (Urban Redevelopment Authority 2003). A total of 44 conservation areas containing over 5,000 buildings had been designated by 2003 and most of these are colonial in origin and character. There is also a Preservation of Monuments Board (PMB) which has gazetted 42 monuments, described as 'enduring historical landmarks' and a 'vital link to the past' (Ministry of Trade and Industry (MTI) 2004), and the majority are again colonial. Popular interest has grown alongside that of the government, evidenced by the setting up of a Heritage Society in the 1980s.

Direct comparisons between Malaysia and Singapore are not always possible due to contrasts in size, geography, resources and political systems with provincial administrations in the former representing a bureaucratic layer not found in Singapore where government is characterised by intense centralisation and control. While there is some legal provision for conservation in Malaysia, it is a 'relatively new practice' (Ghafar Ahmad 1997: 16), with preservation occurring on a limited and ad hoc basis. The federal Department of Museums and Antiquities and the Ministry of Works are becoming more involved, however, and recent anxieties about the destruction of heritage has led to the emergence of non-governmental organisations such as Badan Warisan Malaysia (Heritage of Malaysia Trust) which seeks to 'develop understanding of built heritage as an expression of our history and identity' (Badan Warisan Malaysia (BWM 2004). Sub-national associations such as the Penang Heritage Trust also exist with a mission to protect heritage, seen as one dimension of the 'highly valued social fabric' of Penang state (Penang Heritage Trust 2003).

Ideological, political and social currents therefore underlie the heritage conservation movement, as well as aesthetic sensibilities, but there have also been strong economic motives for official support derived from the notion of heritage as a resource which attracts tourists and their spending (UNESCO 1999). The Singapore authorities displayed their awareness of the revenue earning potential of the past in the 1980s when a report (Wong 1984) claimed that the disappearance of the country's built heritage in the rush towards modernisation and urbanisation was one of the principal causes of a decline in tourist arrivals. This led to a reassessment and the long-term strategy of Singapore Tourism Board (STB), devised when it was still named the Singapore Tourist Promotion Board (Singapore Tourist Promotion Board (STPB) 1996), proposing a core marketing theme which incorporated museums and heritage. The Board is now an enthusiastic advocate of conservation, initiating several projects related to the adaptive reuse of buildings, and the republic's heritage is treated as a marketable commodity. In Malaysia, too, heritage tourism has been encouraged by government as one component of an economic development strategy which extends across the whole service sector (Cartier 1996, 1998).

While there is evidence of a greater commitment to conservation which has helped to rescue some colonial buildings, the extent of this should not be exaggerated and many others have disappeared. There are no comprehensive inventories on which to draw when assessing the scale of the loss, but this is probably considerable and greatest in earlier decades when conservation was even less favoured. There is also some danger of adopting an exclusively commercial orientation in which conservation efforts are frustrated by the over-exploitation of heritage assets. Critics have expressed doubts about the integrity and verisimilitude of certain schemes and the lack of consultation with local stakeholders. As Powell (1994: 27) cautions when writing about Singapore, 'they are directed towards tourists and most of them have profit as the major goal. This has serious implications on the authenticity of the projects and the ability of the conservation areas to retain their former spirit'. Visitor satisfaction thus provides a rationale for conserving buildings, but tourism may also pose a threat to a proper understanding of their historical significance and sympathetic use as explored in the next section.

Colonial heritage as a tourist attraction

An analysis of the colonial townscapes which do survive in Malaysia and Singapore reveals much about perceptions of their histories as colonies. It also illustrates the tourism functions exercised by built heritage and the tensions which can occur when the two phenomena intersect. An earlier study disclosed how remaining properties in the former Straits Settlements serve a variety of purposes, which have often changed over time and acquired a new symbolism, with several appropriated for tourism (Henderson 2002a, 2002b). Simultaneously, several old administrative sites continue to be used by government. Examples include Singapore's neo-classical Supreme Court and City Hall and Georgetown's State Assembly Buildings. However, other government institutions have undergone conversion into heritage attractions such as offices in Singapore which have become the Asian Civilisations Museum and Penang's Fort Cornwallis, now a museum and public park. In Kuching, the Court House is home to the Sarawak Tourism Complex, Fort Margherita houses the Police Museum and the General Post Office is being renovated as a textile museum.

Commercial premises have been subject to mixed fortunes. Some transport termini are still in existence, together with office blocks, while others have seen adaptation and refurbishment. Raffles Hotel in Singapore demonstrates how restoration, reconstruction and adaptive reuse can co-exist, although the outcome may cause confusion (Henderson 2000). It is now difficult to distinguish between parts of the hotel which are original and those added after extensive renovation and refurbishment in the 1980s. The pursuit of revenue is evident in the numerous retail units selling expensive products and 13 restaurants and bars which are part of the complex. Teo and Huang (1995) comment on the alienation of locals, although management maintains that local residents constitute most of the shop, food and beverage customers. The Eastern & Oriental Hotel in Penang, once as famous as Raffles, has also been renovated and its advertising too harks back to a heyday during the colonial era.

Residences for public figures and private individuals and schools have survived, as have religious establishments, but Singapore in particular has witnessed numerous conversions in the uses to which these are now put. Food and beverage and retailing businesses are among the most popular choice for new occupants, exemplified by Chijmes which involved turning the Convent of the Holy Infant Jesus School and its chapel into a network of shops and restaurants. Other social buildings have also mutated, but some retain their earlier role such as the History Museum and Singapore Cricket Club. Eurasians were excluded from the latter and forced to erect separate facilities nearby. The club is therefore a reminder of the divisions and inequalities perpetuated by colonialism, yet it is rarely referred to in this manner by tourism marketers.

Miscellaneous monuments and statues can also be found, erected in tribute to colonial figures, and an esplanade is a landmark of Singapore and Georgetown, as are the botanic gardens founded by the British. Bridges crossing the Singapore River are another obvious legacy, taking their name from prominent notables from the past and of eclectic design, and street names often echo foreign occupation and influences. Georgetown itself was called after the reigning British monarch at its time of assimilation into the Empire.

Colonialism is thus recalled in the urban landscape by the appearance of buildings, their names and sometimes their purposes. However, meanings have altered with use and many of the structures are no longer symbolic of colonial power alone. They have become manifestations of national and state or municipal authority and repositories of heritage, either as museums or indirectly by their very existence. The public and private sectors and local population have recovered these spaces, especially as leisure environments which are enjoyed by tourists and residents. Adaptive reuse is common and facilitates leisure/tourism activities and consumer spending.

It is not only the buildings and their functions which are of relevance to the study of postcolonialism and tourism, but also how these are portrayed in tourist promotional literature. Content analysis of this provides another perspective on the multiple conceptions of colonial heritage. Major sites are included in Singapore's Civic District Trail, a collaborative venture by the National Heritage Board, URA and STB. Participants are offered the chance to discover the 'heart of old Singapore' on a walking route which begins where Raffles reputedly landed in the nineteenth century and proceeds through the Second World War to independence (National Heritage Board 1999). The official Tourist Board Guide (Singapore Tourism Board 2003) also lists these locations under 'Landmarks and Memorials', although references to the colonial period are limited. Buildings are explained principally in terms of current usage or occasions marking progress towards the declaration of the republic and subsequent nationhood. Indeed, there is no mention in the guide that Singapore was a British colony. While Raffles is lauded as the British founder of the port, the brief history then moves on immediately to Lee Kuan Yew, leader of the PAP, and his central role in nation-building.

Georgetown has a heritage trail (Penang Heritage Trust 1999), devised by the local conservation group and supported by the state government. The emphasis is on its multicultural history, with the British only one of many nationalities who settled there. Colonial buildings comprise about a third of those cited and the remainder are Chinese, Indian, Arab and Malay. The actions of Francis Light, who arranged the transfer of the territory to the British in the eighteenth century, is acknowledged and Penang is credited as being part of the 'former British Straits Settlement', but exact dates are not given in the main National Tourism Organisation leaflet (Tourism Malaysia 2001). The authors write generally of a 'fascinating collection of fine old buildings, each bearing the stamp of different foreign influences in its colourful history' and do not highlight the British contribution.

The themes of ethnic and botanical diversity and an eventful and vivid past are prominent in the marketing of Sarawak as a whole (Douglas 1999), with stress on Kuching's 'romantic and unlikely history . . . a hundred-year dynasty of White Rajahs' (Sarawak Tourism Board 2000). The scene of 'heroic adventure and romance, piracy and rebellion' (Tourism Malaysia 2000), Kuching has a legacy of 'remarkable buildings . . . unique examples of colonial period architecture', due to the Brookes, who are praised for their administrative and political skills in a regime that was 'despotic, though benevolent' (Sarawak Tourism Board undated).

The material is open to alternative interpretations and semantic readings regarding the selectivity of the accounts and reactions towards the experience of colonisation. With regard to Singapore, it could be seen as communicating the confidence and pride of a nation looking forward to further attainments. Although the colonial occupation left behind several interesting landmarks, it is now irrelevant except as a theme for tours which promise an 'overview of the trials and tribulations Singapore faced before it became the economic miracle it is today'. The British are no longer a threat and their empire was an historic interlude, long past. There may also be some embarrassment about the period which can be minimised by applying a contextual framework of self-reliance and patriotism to redefine the physical inheritance. Any such uncertainties have not impinged on the marketing of certain enterprises, and Raffles Hotel, for example, capitalises on nostalgia for the age of empire in a language that is mirrored by some Western tour operators. Overall, there is no reluctance to employ the name Raffles, which has been adopted by a shopping mall, the business class on Singapore International Airlines, a private hospital, a marina and golf club as well as streets and schools. Commercially, it is intended to convey an impression of the superior quality of the product or service with connotations of exclusivity and upward mobility.

Less attention is given in the Malaysian literature to the fortunes of Georgetown and Kuching after independence. Georgetown has up to date shopping complexes and hotels and Sarawak's story is that of a 'nation which saw itself make a sudden leap from being a small British colony for a brief period to being one of the more dynamic, progressive and bustling business centres and tourist destinations in Asia-Pacific' (Tourism Malaysia 2000). The contrasting approach may partly be explained by the absence in these locations of the skyscrapers and other modern attributes of Singapore which attest to its recent progress. As in the Singapore narratives, however, the British presence is not dwelt upon and this is perhaps indicative of some ambiguity among Malaysians towards their British colonisers.

Although beyond the scope of this chapter, preliminary observations of museum displays also suggest a degree of ambivalence. Exhibition space is allocated to the formative colonial years, depicted in a straightforward and factual manner, but opportunities are taken to trumpet postcolonial accomplishments and demonstrate national pride. There are no attempts to demonise the British, although aspects of the colonial system are challenged implicitly and explicitly. While not wishing to defend colonialism and while acknowledging the turbulence of the Malayan 'emergency', it should be remembered that the British occupation and withdrawal were not accompanied by the violence and bloodshed apparent elsewhere, and this must be a critical determinant of how colonial powers are judged.

Conclusion

The chapter has raised many issues pertaining to the relationship between heritage and tourism in former colonies that also involve questions of society and culture, political hegemony and economics. It has been argued that heritage in general is increasingly recognised as important in Malaysia and Singapore with growing efforts directed at its conservation, resulting from a combination of imperatives. Some of the colonial built heritage has thus survived and expresses the history of the country and identity of its residents, acting as administrative and business premises and a tourist attraction. In terms of their presentation to domestic and international visitors, the story told by the buildings is not one of subjugation at the hands of an occupying force. Rather, it is a prelude to the successes of the republic in Singapore and a testimony to the vibrant and bustling trading port which was once Georgetown. Kuching's history and the unique reign of a British family set it apart, but the Brookes are hailed as romantic adventurers and celebrated in the buildings they constructed. Attention is also drawn to the visual appeal of colonial architecture, often enhanced by its site and setting. The relationship with former colonial powers in the postcolonial world of tourism would thus no longer seem to be one of subservience, but whether it is yet that of equal partners is a topic for continued debate.

References

Ashworth, G. J. (2000) 'Heritage, tourism and places: a review', *Tourism Recreation Research* 25, 1, 19–29.
Baker, J. (1999) *Crossroads: A Popular History of Malaysia and Singapore*, Singapore: Times Books International.
Badan Warisan Malaysia (BWM) (2004) Badan Warisan Malaysia, www. badanwarisan.org.my, accessed 1 June.
Beamish, J. and Ferguson, J. (1985) *A History of Singapore Architecture: The Making of a City*, Singapore: G. Brash.
Britton, S. G. (1982) 'The political economy of tourism in the Third World', *Annals of Tourism Research* 9, 331–58.

Cartier, C. (1996) 'Conserving the built environment and generating heritage tourism in Peninsular Malaysia', *Tourism Recreation Research* 21, 1, 45–53.

Cartier, C. (1997) 'The dead, place/space, and social activism: constructing the nation-scape in historic Melaka', *Environment and Planning D: Society and Space* 15, 5, 505–632.

Cartier, C. (1998) 'Megadevelopments in Malaysia: from heritage landscapes to leisurescapes in Melaka's tourism sector', *Singapore Journal of Tropical Geography* 19, 2, 151–76.

Chew, E. C. and Lee, E. (eds) (1991) *A History of Singapore*, London: Oxford University Press.

Crouch, H. A. (2001) 'Managing ethnic tensions through affirmative action: the Malaysian experience', in N. J. Colletta, T. G. Lim and A. Kelles-Viitanen (eds) *Social Cohesion and Conflict Prevention in Asia: Managing Diversity through Development*, Washington, DC: The World Bank, pp. 225–62.

Dale, O. J. (1999) *Urban Planning in Singapore: The Transformation of a City*, Kuala Lumpur: Oxford University Press.

Douglas, N. (1999) 'Towards a history of tourism in Sarawak', *Asia Pacific Journal of Tourism Research* 3, 2, 14–23.

Edwards, N. and Keys, P. (1988) *Singapore: A Guide to Buildings, Streets, Places*, Singapore: Times Books International.

Ghafar Ahmad, A. (1997) *British Colonial Architecture in Malaysia 1800–1930*. Kuala Lumpur: Museums Association of Malaysia.

Graham, B., Ashworth, G. J. and Tunbridge, J. E. (2000) *A Geography of Heritage: Power, Culture and Economy*, London: Arnold.

Gurnstein, P. (1985) *Malaysian Architectural Heritage Survey: A Handbook*, Kuala Lumpur: Heritage of Malaysia Trust.

Hayes Hoyt, S. (1991) *Old Penang*, Kuala Lumpur: Oxford University Press.

Hayes Hoyt, S. (1993) *Old Malacca*, Kuala Lumpur: Oxford University Press.

Henderson, J. C. (2000) 'Conserving colonial heritage: Raffles Hotel in Singapore', *International Journal of Heritage Studies* 7, 1, 7–24.

Henderson, J. C. (2002a) 'Tourism and colonial heritage in Singapore', *Tourism, Culture and Communication* 3, 3, 117–29.

Henderson, J. C. (2002b) 'Built heritage and colonial cities', *Annals of Tourism Research* 29, 1, 254–57.

Hill, M. and Lian, K. F. (1995) *The Politics of Nation Building and Citizenship in Singapore*, London: Routledge.

Jacobs, J. M. (1996) *Edge of Empire: Postcolonialism and the City*, London: Routledge.

Jayapal, M. (1992) *Old Singapore*, Singapore: Oxford University Press.

King, A. D. (1976) *Colonial Urban Development: Culture, Social Power and Environment*, London: Routledge and Kegan Paul.

Liu, G. (1996) *In Granite and Chinam: The National Monuments of Singapore*, Singapore: Landmark Books.

Ministry of Trade and Industry (MTI) (2004) Ministry of Trade and Industry. www.mti.gov.sg, accessed 1 June.

Morgan, N. and Pritchard, A. (1998) *Tourism Promotion and Power: Creating Images, Creating Identities*, Chichester: John Wiley.

National Heritage Board (1999) *Civic District Trail: Rediscover the Heart of Singapore*, Sydney: Z-CARD Asia Pacific.

Palmer, C. (1994) 'Tourism and colonialism: the experience of the Bahamas', *Annals of Tourism Research* 21, 4, 792–811.

Palmer, C. (1999) 'Tourism and the symbols of identity', *Tourism Management* 20, 313–21.

Payne, R. (1986) *The White Rajahs of Sarawak*, Singapore: Oxford University Press.

Penang Heritage Trust (1999) *Penang Heritage Trail: Exploring the Streets of Georgetown*, Penang: Penang Heritage Trust.

Penang Heritage Trust (2003) Penang Heritage Trust, www.pht.org.my, accessed 14 June.

Porritt, V.L. (1997) *British Colonial Rule in Sarawak*, Kuala Lumpur: Oxford University Press.

Powell, R. (1994) *Living Legacy: Singapore's Architectural Heritage Renewed*, Singapore: Singapore Heritage Society.

Rodan, G. (1996) 'Class transformations and political tensions in Singapore's development', in R. Robison and D. Goodman (eds) *The New Rich in Asia: Mobile Phones, McDonald's and Middle Class Revolution*, London: Routledge, pp. 19–45.

Said, E. W. (1994) *Orientalism*, New York: Random House.

Sarawak Tourism Board (2000) *Kuching and Southwest Sarawak: The Hidden Paradise of Borneo*, Kuching: Sarawak Tourism Board.

Sarawak Tourism Board (undated) *Jejak Warisan: Kuching Heritage Trail*, Kuching: Sarawak Tourism Board.

Shamsul, A. B. (1999) 'From Orang Kaya Baru to Melayu Baru: cultural construction of the Malay "new rich"', in M. Pinches (ed.) *Culture and Privilege in Capitalist Asia*. London: Routledge, pp. 86–110.

Shaw, B. J. and Jones, R. (1997) *Contested Urban Heritage: Voices from the Periphery*. Aldershot: Ashgate.

Simon, D. (1984) 'Third World colonial cities in context', *Progress in Human Geography* 8, 4, 493–514.

Singapore Tourism Board (STB) (2003) *Official Guide Singapore New Asia*, Singapore: Singapore Tourism Board.

Singapore Tourist Promotion Board (STPB) (1996) *Tourism 21 Vision of a Tourism Capital*, Singapore: Singapore Tourist Promotion Board.

Smith, R. A. (1999) 'Urban Redevelopment Authority: tourism and heritage conservation in Singapore', in D. H. Hooi (ed.) *Cases in Singapore Hospitality and Tourism Management*, Singapore: Prentice Hall, pp. 39–57.

Southall, A. (1971) 'The impact of imperialism upon urban development in Africa', in V. Turner (ed.) *Colonialism in Africa, 1870–1960*, London: Cambridge University Press, pp. 216–54.

Tarling, N. (1971) *Britain, the Brookes and Brunei*, Kuala Lumpur and New York: Oxford University Press.

Teo, P. and Huang, S. (1995) 'Tourism and heritage conservation in Singapore', *Annals of Tourism Research* 2, 1, 589–615.

Tourism Malaysia (2000) *Sarawak, Malaysia: The Best of Borneo*, Kuala Lumpur: Tourism Malaysia.

Tourism Malaysia (2001) *Penang*, Kuala Lumpur: Tourism Malaysia.

Tunbridge, J.E. and Ashworth, G.J. (1996) *Dissonant Heritage: The Management of the Past as a Resource in Conflict*. Chichester: John Wiley.

Turnbull, C.M. (1989) *A History of Malaysia, Singapore and Brunei*, Sydney: Allen and Unwin.

UNESCO (1999) *The Economics of Heritage. Conference/Workshop on the Adaptive Reuse of Historic Properties in Asia and the Pacific*, 9–17 May, Conference Programme, UNESCO.

Urban Redevelopment Authority (URA) (Singapore) (2003) *Concept Plan 2001*, URA Online. www.ura.gov.sg, accessed 27 January.

Vlatseas, S. (1990) *A History of Malaysian Architecture*, Singapore: Longman.

Watson, C. W. (1996) 'The construction of the post-colonial subject in Malaysia', in S. Tonnesson and H. Antlov (eds) *Asian Forms of the Nation*, London: Curzon Press, pp. 270–96.

Western, J. (1985) 'Undoing the colonial city?' *The Geographical Review* 75, 3, 335–57.

Williams, P. and Chrisman, L. (1994) *Colonial Discourse and Post-Colonial Theory: A Reader*, Hemel Hempstead: Harvester Wheatsheaf.

Wong, K. C. (1984) *Report of the Tourism Task Force*, Singapore: Ministry of Trade and Industry.

Yeoh, B. (1996) *Contesting Space: Power Relations and the Urban Built Environment in Colonial Singapore*, Kuala Lumpur: Oxford University Press.

8   A colonial town for neocolonial tourism

David Fisher

Introduction

The hypothesis that tourism can be a form of neocolonialism has been suggested by a number of writers (for example, Nash 1977; Britton 1982; Hall 1994). Tourism in developing countries can enhance the economic power of a tourist generating region over a destination region, thereby enabling foreign interests to dictate development programmes (Bull 1995; Wilkinson 1997). Colonialism and imperialism as historic phenomena are also of interest to tourism scholars, especially in some aspects of heritage tourism. The relics of imperialism can act as an attractor for tourism. Many countries that were once controlled by imperial powers have ruins, arte-facts and cultural remnants of those powers that are of interest to tourists. For example, many destinations around the Mediterranean and in Europe encourage visitors to see Roman ruins. Reconstructions of colonial towns, such as Jamestown in the United States, have been created to give the descendants of the Imperial power a link with their past. Imperial history, in the form of heritage tourism, has become an expanding tourist market.

When combined, tourists visiting formally colonised countries because of an interest in imperial history can give rise to a new type of conflict. The historic relics of an imperial power in a colonised state may encourage a form of tourism that is fundamentally neocolonialist. Spatial conflicts resulting from tourism are well documented (for example, de Kadt 1979; Woodcock and France 1994; Din 1997; Wilson 1997). Competition may occur between tourism entrepreneurs and local people for the use of land and other resources. However, heritage tourism, particularly heritage tourism that takes place in a cross-cultural environment, can result in temporal conflict – conflict over the meaning of the past. Questions can be raised about whose heritage is being portrayed and whose heritage is being written or re-written. The meaning of place for all the actors involved may be altered because of heritage tourism.

While cultural and heritage tourism are becoming increasingly important (Alzua *et al.* 1998), the contested nature of natural and built heritage is not a topic that has been examined in depth in the tourism literature.

However, as Morphy (1993: 206) points out when commenting on Australia, '(l)andscape provides an excellent framework for representing the clash in values and the different interests of Aborigines and colonists'. Landscapes, including the built environment, can have a variety of cultural interpretations resulting in contested space and history (Kirby 1993). The value of a place may then change as different groups are influenced by each other as well as by exogenous factors. Baker (1992: 3) notes that 'landscapes are shaped by mental attitudes and that a proper understanding of landscapes must rest upon the historical recovery of ideologies'. It is the ideology that reflects the values placed on aspects of the environment in which people live. Neocolonialism may bring with it an alternative ideology that can affect the values of the local community.

Tourism can have considerable impact on the meaning of place because, while tourists may only visit for very short periods, their collective impact in terms of numbers and income to the community may be enormous. In addition, tourists' understanding of local cultures will remain superficial, given the relatively short duration of their stay. If viewed from the perspective of the hosts, a succession of people will appear to have little knowledge of the destination culture but a strong sense of their own values, which may conflict with those of the hosts.

This chapter examines Levuka, the old capital of Fiji, on the island of Ovalau, a destination marketed for its heritage and colonial attributes. Levuka was created and built by colonialists in a European style but largely abandoned by them before colonial rule ended. It is now an attraction for 'European' visitors. (In Fiji, the term European is used to mean people who are ethnically European, thus including Australians, New Zealanders and North Americans. Sometimes people of other races are included in the European grouping for simplicity. For example, a native American in Levuka was classed as European.) This chapter demonstrates how tourists and recent non-indigenous residents impose their heritage values on the host community, and how heritage is a distinctly cultural phenomenon. This chapter is the result of eight visits to Levuka between 1984 and 2000, including one lasting ten months in 1996–7.

Historical identity, and the heritage that goes with it, is constructed through both time and culture (Olwig 1999). Heritage, in its broad sense is an etic concept, that is, it is a concept that is common across cultures. However, what constitutes heritage is an emic concept, or is specific to particular cultures, places and time. The question is whether changes in the emic view are, in fact, imposed from other cultures. Is tourism a neocolonialist means by which the symbols of the culture of the destination change in response to what visitors value?

Levuka illustrates the conflicts that colonial heritage tourism can generate and shows that, while the colonial power may have gone, the economic power of neocolonialism can attempt to dictate the development of a destination in a way that would not occur if there was no colonial legacy.

Levuka: a colonial town

In the early nineteenth century the Fiji islands were made up of a number of chieftainships, which vied for local political power through marriage and warfare. It was not until the arrival of European weapons technology that small tribal factions began to coalesce into larger units. In 1870 the first attempt at a unified nation of the Fiji islands was attempted by Tui Cakobau who styled himself King of Fiji and created a government made up of many European traders and farmers. However, this government did not gain universal approval because of Tongan expansionism from the east and European opponents in the main town and *de facto* capital of Levuka, a European settlement of about 1,000 people. Increasing levels of lawlessness led Cakobau to cede Fiji to Queen Victoria in 1874. Levuka then became the official capital of the new colony. (For detailed accounts of nineteenth-century Fiji see Derrick 1950; Scarr 1984; and Routledge 1985.)

Levuka itself was founded by an American sailor, David Whippy, in the 1820s when he was given land by the local chief, Tui Levuka. It soon became the centre of European trade in the Fiji Island group and a distinctly European town was created next to the Fijian village of Levuka Vakaviti. As more Europeans arrived European style housing was built along the waterfront with hotels, bars and businesses vying for the limited flat coastal land. It became the first town in Fiji, and in the South Pacific outside Australia and New Zealand, to have banks, newspapers and electricity. Commerce was centred on trade with other colonial states, Europe and the US. It developed as a distinct European colonial town in the South Pacific, markedly different in style and structure from Fijian villages.

In 1882 the British colonial administration moved the capital to Suva, on the main island of Viti Levu. Steeply rising hills behind the town and the sea in front prevented Levuka expanding. In addition, development in shipping technology required a deeper harbour than was available at Levuka (Young 1993). For a while Levuka maintained its position as an important commercial centre but as Suva began to dominate the political life of the colony, commerce gradually shifted to the main island. By the end of the nineteenth century most of the available land had been used for building and little incentive existed to replace older buildings with more modern ones. Business continued in the structures that had been constructed in the heyday of the town. Since 1904 only a couple of concrete buildings have been erected in the centre of the town, and on the outskirts a tuna freezing works and canning factory was built on reclaimed land at the start of the 1960s.

As a result of the geo-political changes in Fiji, at the start of the twenty-first century Levuka has been left as a little-changed nineteenth-century colonial town. In 1907 the *Cyclopedia of Fiji* (Allen 1984 (1907): 263) suggested that '(e)very visitor to Levuka is charmed with its delightful surroundings'. This is true of tourists today. The *Lonely Planet Guide* for the South Pacific (Jones and Pinheiro 2000: 548) describes Levuka as 'a slow-paced, picturesque place with buildings reminiscent of a Wild West tumbleweed town'.

Heritage in Levuka

There is more than one cultural definition of heritage in Levuka. The different ethnic groups all have a different understanding of conservation and the meaning of place. Very loosely, history and heritage for Fijians is closely linked with the concept of *vanua,* that is, land and the cultural and genealogical attachments to it. For Europeans it is more closely linked with objects, which include buildings. Europeans, especially tourists, see their history, culture and heritage by what is on the land. For the Indo-Fijians, any meaning of place is based around family history and not on the history of the area.

The fact that few new buildings were erected in the twentieth century in the main thoroughfare of Beach Street has been due to an uncertain economic future for the town, and a lack of space. The existing buildings have been partially maintained in a climate of 'benign neglect' (Samudio 1996) but not updated to any great extent. This attitude is changing in some quarters. Levuka has been 'discovered' by sections of the tourist industry and a move from functional maintenance to conservational maintenance is being encouraged. The former is concerned with maintaining a building so that it can be used to carry out the activity required of it in the most cost efficient way. Any maintenance is not concerned with the historical integrity of the building. This is in contrast to conservational maintenance, which is wholly concerned with the preserving the building in a particular form.

Building preservation

The belief in the need to preserve Levuka as a colonial town has come from resident Europeans and outside agencies, who appear to view the conservation from their own perspective without considering that non-European residents may have different views on the intrinsic value of the old buildings. The organisations charged with investigating the heritage value of Levuka have come with a primarily 'Western' view of what constitutes heritage.

The first documented indication of a desire to preserve the architecture of the town was the Belt Collins and Associates report of 1973:

> In contrast to the other Accommodation Regions [*sic*], which are based primarily on attractions of the natural environment, the town of Levuka also offers attractions of great historical interest as reflected in the unspoiled nineteenth century building styles. Levuka, with its history as a lively mid-nineteenth century port and whaling centre, the former

capital of Fiji, and location of the Deed of Cession to England [*sic*], combined with its charming hundred year old store fronts and hillside houses, accessible only by pedestrian paths overlooking the harbour, can become a major point of interest in the form of day trips from Suva as well as overnight visits. The keys to achieving this are historical preservation of the town and improved transportation from Suva.

(1973: 70)

In 1977 the Levuka Historical and Cultural Society (LHCS) was founded. This is a local body which, in 1997, consisted of residents and holiday home owners, none of whom were ethnic Fijians apart from those who held honorary positions and seldom took part in meetings. The object of the society is to maintain the historic colonial architecture of the town.

In the same year as the founding of the LHCS, a further report based on the findings of the initial Belt Collins report was issued by the Fiji Visitors' Bureau. This stated that work on the preservation and restoration of Levuka should follow three stages: first aid; preservation; and restoration. This document considers the practical problems of preservation and the role of the government and the public sector. It also comments favourably on the preservation of Lahaina, the old capital of Hawai'i stating that 'much of Lahaina's charm and historical interest has been retained' (Belt Collins and Associates Ltd 1977) and that Lahaina is now a leading tourist destination in Hawai'i. The interest in Lahaina comes from the fact that it is a town with which Levuka is twinned.

The Pacific Area (now Asia) Travel Association (PATA) Task Force carried out the first detailed analysis of heritage buildings in Levuka. Geoffrey Bawa of Sri Lanka, who received the PATA 'Heritage Award of Recognition' for his work on the Sri Lankan Parliamentary Complex, organised this analysis. He 'has had extensive experience in the design of international tourist resorts and hotels' (Pacific Area Travel Association 1985: 39). The analysis and resulting report on the heritage of the buildings is solely attributed to him.

Bawa points out that while many of the buildings, both old and new, are of no great interest individually, 'it is the totality of the place, the continuity of new and old, which captures the eye and would delight any visitor' (Pacific Area Travel Association 1985: 39). The effect of this totality is that 'unharmonious [*sic*] changes in the face of any single building constitute a danger to, if not destruction of, the value of the whole long beach-front elevation' (1985: 39). Bawa cannot understand why there had been no conservation and tourism development in the period following the Belt Collins report and the visit of the PATA task force. He makes it very clear what his feelings are towards Levuka.

It is widely recognised that Levuka is lovely: it evokes love at first sight.

> Its history is well known, the setting is stunning, the visual elements are inspiring, the tourism potential is obvious, the planning solutions are simple, the government participation is justifiable, the people are warm, the council is willing, and the need for action is compelling!
>
> (Pacific Area Travel Association 1985: 43)

He sees an obvious link between the town's future economic prosperity and the development of tourism through the preservation of the town's heritage, though he too states that Levuka is 'the seat of Fiji's history' (Pacific Area Travel Association 1985: 43) and that it 'represents an important stage in the history of the Pacific in its manifestation of nineteenth-century Western architecture. It holds great historic associations for the people of Fiji and demonstrates visually a phase of development of the nation' (1985: 39). However, no evidence is provided to support the view that that is how the people of Fiji view the town. Levuka is not considered important in the history of Fiji by Fijian islanders other than those of European descent (personal communication by a Fijian journalist visiting Levuka).

As a result of the PATA task force the government of Fiji designated Levuka a Heritage Town in 1990, and in 1994 PATA financed the appointment of a heritage adviser for the town (though one member of the task force claims that he personally paid the adviser's salary as PATA did not have the funds to do this).

Also in 1994 the Department of Town and Country Planning and PATA commissioned a report by HJM Consultants Pty Ltd and Timothy Hubbard Pty Ltd (known as the Hubbard Report). The methodology for the study was based on the 1990 New South Wales Conservation Plan. The Hubbard report stated that in Levuka and the surrounding areas on Ovalau over 120 places had been identified as being historically significant. A number of recommendations were made that the government had moral and political difficulty in accepting (personal communication by John Bennett, heritage adviser 1997). These included the confiscation of all land for which rates had not been paid for a number of years. There is a chronic low incidence of rates payments in Levuka. In 1997–8 no rates were received from any household in the suburb of Baba nor had there been for a number of years. It is now being suggested that the town boundaries are reduced so that Baba is no longer included within them (personal communication by David Kirtin, former heritage adviser 1999). (This is in direct opposition to the report's later recommendations that the town boundaries be extended. However, their rationale is to put all historic buildings and sites, such as St John's Church and the Draiba cemetery under council jurisdiction. There is nothing of historic significance in Baba, which is a peri-urban site without proper sanitation or electricity supply.) The Hubbard Report (HJM Consultants Pty Ltd and Timothy Hubbard Pty Ltd 1994: 4) goes on to say that following confiscation 'the council would lease these properties for appropriate new buildings and uses which would encourage additional business to the town such as tourism'.

The extension of the town boundaries would be a prelude to putting control of the protected areas of Levuka and Ovalau under one authority and preventing the uncontrolled spread of both development and jurisdiction of heritage. At a much later point in the report it is acknowledged that there may be some problems with implementing this recommendation as land would have to be taken from the Native Land Trust Board (NLTB). There would also have to be a 're-examination of the extent of coastal controls' (HJM Consultants Pty Ltd and Timothy Hubbard Pty Ltd 1994: 40). The NLTB controls the land set aside by the British colonial administration as land that can only be owned by Fijians. It would be extremely unlikely that any government would alter this fundamental structure of land ownership anywhere in the country. Its importance is reflected in the argument that one of the reasons for the 1986 coup was fear that the new government would do this (Ravuvu 1991). Interestingly, a similar fear was expressed following the election of the Indian dominated Labour government that was overthrown in 2000 by George Speight.

The final point made in the PATA report was that in 1991 Fiji became a signatory to the World Heritage convention. If the Fijian government were to nominate any places for the World Heritage List 'it would have to show that it has legislative measures in place to properly protect such places' (Hubbard HJM Consultants Pty Ltd and Timothy Hubbard Pty Ltd 1994: 60). In the opinion of the writers of the report Fiji did not have any heritage protection legislation nor did it have good wildlife protection. By 1997 heritage protection in Fiji was still inadequate and there was little chance of the town boundaries being extended (personal communication by John Bennett, heritage adviser 1997).

The purpose of buildings

There are three distinct groups among the local residents: the ethnic Fijians, of whom few live in the historic parts of the town; the Indo-Fijians and Chinese shopkeepers, who own most of the shops in Beach Street; and Europeans made up of the 'old' European families, people descended from the colonial settlers, and the recent 'European' arrivals, who set up homes in the town because of its historic ambiance. For each group the town has a different meaning.

The Fijian attitude to heritage conservation is complex. Some people see no point in it at all and think that preserving the buildings detracts from the town to such an extent that it discourages potential businesses and tourists from visiting. Others believe that if the buildings are preserved they will attract tourists, bringing income and providing employment. However, this group would not maintain the buildings if tourists do not visit. Their concern for preservation is entirely dependent on the income it can generate. Some believe that the buildings are preserved for tourists but that this is a good thing because it keeps an aspect of Fijian history alive. Finally, a minority argue that the buildings should be kept for their own sake; that preservation is not and should not depend on tourists.

The reasons for these differing views appear to be based on functionality. Many non-European local residents believed that the buildings should be knocked down because modern buildings are more functional. Very few considered the buildings to be representative of Fijian history. There are two reasons for this. The first is related to the Fijian conceptualisation of history and the second is concerned with what Fijians feel are the important moments in their history.

The best example of the first point came from a female respondent. In answer to the question, 'Why do you think that tourists come to Levuka?', she replied that it was because of the history. Later in the interview she was asked, 'Do you think the old buildings and monuments in Levuka should be preserved?' She replied that they should be pulled down and replaced with new ones. She had also said that she wanted more tourists to visit Levuka. Sensing a contradiction here I reminded her of this. She looked at me with a markedly surprised expression and said, 'The history is still here!' She explicitly stated what many other respondents assumed that I understood: that history is not to be found in the buildings but in the land, location and *vanua*. A sense of *mana* is passed from an institution that uses a building, such as a priest or chief's bure, but it is the institution and not the building that has that *mana*. Once the building has gone the place where it was maintains the *mana* even if an alternative building has been built elsewhere. From a Western tourist's point of view this can be confusing.

The second point that is necessary in order to understand the Fijian attitude towards Levuka's built environment is the conceptualisation of the history that created the town in the first place. Levuka Town (as opposed to Levuka Vakaviti) was created by Europeans and is still seen as a European town. It is not Fijian. As was noted earlier, arguments put forward for the preservation of the town included the belief that the town is the birthplace of the nation. While this may be true in the sense of the modern nation state Fijians look beyond that. One respondent said that this is not of much concern to most Fijians. They see their origins in the legends of their ancestors' arrival in Fiji and, more importantly, in the area that their clan or kin group originated. These stories are still told on Ovalau. The people of Lovoni, a village in the centre of Ovalau, trace themselves back to the first person to arrive on the island, who swam there to escape persecution on Viti Levu. A prominent nationalist politician was reported as saying that Levuka is not part of Fijian heritage and therefore is of no value to Fijians (personal communication by ex heritage adviser 2000).

Another respondent put the problem of the preservation of the town very succinctly from a Fijian point of view. She did not understand why Fijians should have to pay to preserve a European town in a European way for Europeans (i.e. the tourists).

The attitudes of the European residents of Levuka are very different. While they come from a variety of backgrounds, most feel that it is their duty to give back something to the town. Heritage is one aspect of this but it is also relevant to other areas of voluntary work.

Most of the European residents are involved with community projects and organisations such as the LHCS and the creation of a school for the disabled. As one stated 'I became soon well aware of my obligations in regard to a bit of energy expended but there was obvious need and a lot of interest in being active within the community and I was able to give time to the Levuka Historical Society' (personal communication 1997). Nevertheless, this same respondent also stated that if no benefits are accrued to the community centre from his efforts, due to inefficiency, inertia or political infighting of the committee, 'we'll do it for some other charity'. He is on the board of the Levuka Society for the Disabled. Another of the European residents stated that he is willing to provide his services but:

[o]ne thing I deplore and absolutely detest is going round asking people for money. I'd rather do it myself and do voluntary things and do things for nothing rather than go and ask for people to pay me to do it or ask for money from people.

(Personal communication 1997)

Typical of the new European residents' responses to questions on the validity of heritage preservation in the town are: 'Well it's . . . history, it's the historical landmark of the origins of the . . . town. You take away the landmarks of the town you take away the heritage, the culture from the town and they (people generally) have no specific thing to look at. Nothing material there to see.'

Another European resident stated that while she liked the idea of preservation she did not like tourists. Levuka attracted her because of its ambience. This was created by the architecture of the town. However, that would disappear if the buildings went or if too many people came to see them. As another European resident stated: 'It seems that there is a mixture of visions for Levuka. The town's people seem to be letting the outsiders fight it out.'

The shopkeepers have mixed views on heritage preservation. Those who run shops that cater for tourists are generally more in favour of the heritage legislation. One respondent was concerned that all the shopkeepers had to cover the costs of preserving the heritage while it was only the tourist businesses that made the money; tourists bought very little in this respondent's shop. He also complained that the heritage rules did not allow for sensible alterations. Another shopkeeper did not believe that the council was interested in helping shopkeepers. This may be because the only councillors who had a stake in Beach Street both worked in the tourism sector. A third respondent felt that historic preservation was geared against Indian shop-keepers. This was an extreme view of the feeling of some shopkeepers that outsiders controlled heritage conservation. One stated that rules were regularly broken and no action was taken. He inferred that if alterations were done quietly, no one minded; however, this prevented major alter-ations. One story that was repeated on a number of occasions concerned a resident who was told that he could not replace louvre windows with new ones. He was told that he had to put in windows of a traditional design. However, these cost a lot more. Eventually a compromise was reached whereby he was able to use louvre windows made of wood. Nevertheless, the whole process took over a year before agreement was reached.

Major alterations to buildings are not allowed so requests by shopkeepers to extend their buildings have been refused. One shopkeeper explained that providing work and a home for family members was part of Indian culture; the building within which the enterprise took place was only a component of the business and not the reason for it. Most respondents were happy to preserve the façade of the buildings but wanted to be able to change other parts of them so that they could function more efficiently as the site of the business and a place for the family. The buildings were not objects in their own right but integral parts of the business. To have one without the other was meaningless.

Conclusion

Levuka illustrates a variety of interpretations on what constitutes the value of historic buildings. The attempts by outsiders to maintain the existing townscape for aesthetic and cultural reasons is being promoted to local people as a means of providing work and income. The hope is that 'European' heritage values will be followed by the local people at a later stage. Tourism is seen as the means by which the money needed for the preservation can be obtained. The income generating possibilities of colonial heritage is also used as a way to persuade local people that preservation is of value to them.

Nevertheless, there is a lack of mutual understanding of the terms that have been used in promoting tourism in Levuka. A variety of meanings are given to terms such as heritage, tourism and tourists without a reali-sation that other groups interpret the words differently. The only unifying factor is the hope that tourism will increase income in the town. What most local Fijians would like is a regular source of income. It is becoming increasingly difficult for them to maintain a subsistence lifestyle as they are subject to continuous demands for cash from the villages, churches and schools. Income from working at the canning factory is not guaranteed as the factory shuts down intermittently. This affects not only the workers and their families but also the shopkeepers who rely on their income.

Of the new European residents in the town, all but three originally visited either as tourists or because of involvement in a tourist enterprise. It is these people who have been the driving force for the historical preservation of the town. Two of the Europeans involved in the heritage of the town accept fully that the preservation of the built heritage is something that has been demanded by outsiders. They justify this by saying that the motto for local people is 'you take the lead'. These two were adamant that where they led was where the local people wanted to go.

From the point of view of many local people, both Indian and Fijian, it appears that the town has been picked out by travel agents as a good place to 'market'. This idea has been taken up by other outsiders who have attempted to enforce their own beliefs in conservation. However, there is disgruntlement because it is not the outsiders who are expected to pay the financial costs of the heritage preservation but the local owners of the properties. There are also a number of people that are sceptical about the benefits that will accrue to them from increased tourism, especially in areas away from Levuka Town. People in Levuka Vakaviti believe that those involved in tourism in the town want to keep all the money to themselves and not spread it around the greater community. This view was also commented on in a government report on tourism in Ovalau (Nawadra 1995).

Many local people have been persuaded that conservation will attract tourists and, although many local people say they enjoy contact with tourists, they predominantly want the income that they think that tourists will provide. The demand for this tourism-derived income has consequently led to the desire to preserve the town. There is no cultural understanding of why tourists want to see old buildings (though some think that the tourists want to see their own cultural heritage). However, if tourists do want to see them and provide work and income for local people as a consequence, then many local people are happy to keep the buildings. Others tend to view what tourists want from their own perspective. They want to show Fiji as a modern, progressive country and feel that this is what tourists want. 'Tourists want modern things', said one respondent. Another respondent, who had received higher education, claimed that many Fijians are ashamed of their own culture and want to appear modern and progressive to the outside world. For them the idea of preserving the old is incomprehensible and to preserve Levuka is to tell the world (i.e. tourists) that Fiji has not developed in the past hundred years.

Is tourism changing the way local people value and construct the meaning of heritage? The answer to this is not straightforward because it is possible for people to behave in similar ways but for different reasons. The majority of local residents are in favour of preservation *if* it encourages tourism – so by that definition tourism has changed attitudes or beliefs. However, this change only occurs if tourism increases. Should tourism not increase as a result of building preservation then there is no point in preserving the buildings. The fact that tourists value the old buildings for their own intrinsic quality has not resulted in many local people valuing them in the same way – the value of the buildings for local people is derived from their financial value. More importantly, there is very little evidence of a change in behaviour among the local people: very few local people are involved in the physical conservation of the town unless they own a historic building and then it tends to be a grudging involvement.

In addition, at a culturally deep level is the concept of *vanua*. Land and the physical environment have far greater meaning to Fijians than to Europeans. Conceptually, *vanua* encompasses spiritual values that include the past and many aspects of heritage. For Europeans it is the buildings that explain the past, which is why Levuka is promoted as a tourist destination by local Europeans. There may be signs that Fijians with higher levels of education who come from outside Levuka appreciate the 'European' values of preserving the built environment but there is not adequate evidence to make a definitive statement.

What is clearly apparent is that there are a number of different worldviews in existence at the same time. If people who hold a particular worldview do not acknowledge that other people may have a different conceptualisation of the world, or the validity of these different conceptualisations, then misunderstandings and hostility are likely to occur. Buildings are being preserved in Levuka for the following reasons: they look nice; they are historical; the owners cannot afford to change them; owners believe that they are not allowed to change them; they bring in tourists and therefore money; the government says they must be preserved; foreigners say they must be preserved. However, there is little appreciation of the differences in meaning that are attached to buildings.

In the case of Levuka the entwinement of worldviews is one way. It is local people who have to accommodate the worldview of outsiders. While this does not require local people to give up their own worldviews, they have to see the preservation of the town in terms of income generation in order to support it. Outsiders have just assumed that they are 'correct' in preserving the town and that the local people will come round to their way of thinking some time in the future. There appears to be no attempt to accommodate the worldviews of local people or even to consider the possibility that different worldviews exist. In this sense tourism can be seen to be an agent of neocolonialism. Whether it is a successful agent remains to be seen.

References

Allen, P. S. (1984 (1907)) *Cyclopedia of Fiji*, Suva: Fiji Museum.

Alzua, A., O'Leary, J. T. and Morrison, A. M. (1998) 'Cultural and heritage tourism: identifying niches for international travellers', *Journal of Tourism Studies* 9, 2: 2–13.

Baker, A. R. (1992) 'Introduction: on ideology and landscape', in A. R. Baker and G. Biger (eds) *Ideology and Landscape in Historical Perspective: Essays on the Meanings of Some Places in the Past*, Cambridge: Cambridge University Press, pp. 1–14.

Belt Collins and Associates (1973) *Tourism Development Programme for Fiji*, United Nations Development Program.

Britton, S. (1982) 'The political economy of tourism in the third world', *Annals of Tourism Research* 3: 331–58.

Bull, A. (1995) *The Economics of Travel and Tourism* (2nd edn), Melbourne: Longman.

de Kadt, I. (1979) *Tourism: Passport to Development?*, New York: Oxford University Press.

Derrick, R. A. (1950) *A History of Fiji*, Vol. 1, Suva: Printing and Stationery Department.

Din, K. H (1997) 'Tourism development: still in search of a more equitable mode of local involvement', in C. Cooper and S. Wanhill (eds) *Tourism Development: Environmental and Community Issues*, Chichester: Wiley, pp. 153–62.

Hall, C. M. (1994) 'Ecotourism in Australia, New Zealand and the South Pacific: appropriate tourism or a new form of ecological imperialism?' in E. Cater and G. Lowman (eds) Ecotourism: A Sustainable Option? Chichester: John Wiley & Sons, pp. 137–58

HJM Consultants Pty, and Timothy Hubbard Pty (The Hubbard report) (1994) *Levuka Heritage and Conservation Study for Department of Town and Country Planning and PATA*, unpublished report.

Jones, R. and Pinheiro, L. (2000) *Fiji*, 5th edn, Melbourne: Lonely Planet.

Kirby, V. G. (1993) 'Landscape, heritage, and identity: stories from the West Coast', in C. M. Hall and S. McArthur (eds) *Heritage Management in New Zealand and Australia*, Auckland: Oxford University Press, pp. 119–29.

Morphy, H. (1993) 'Colonialism, history and the construction of place: the politics of landscape in Northern Australia', in B. Bender (ed.) *Landscape: Politics and Perspectives*, Providence: Berg, pp. 205–44.

Nash, D. (1977) 'Tourism as a form of imperialism', in V. L. Smith (ed.) *Hosts and Guests: The Anthropology of Tourism*, Philadelphia: University of Pennsylvania Press, pp. 33–47.

Nawadra, S. (1995) *Ovalau Integrated Management Plan*, Project Issues Paper.

Olwig, K.F. (1999) 'The burden of heritage: claiming a place for a West Indian culture', *American Anthropologist*, 26, 2: 370–88.

Pacific Area Travel Association (PATA) (1985) *Levuka and Ovalau. Tourism Development through Community Restoration*, Sydney: PATA.

Ravuvu, A. (1991) *The Facade of Democracy*, Suva: Reader Publishing House.

Routledge, D. (1985) *Matanitu: The Struggle for Power in Early Fiji*, Suva: University of the South Pacific.

Samudio, J. (1996) *Summary of Heritage Activities: Levuka, Fiji*, unpublished report.

Scarr, D. (1984) *Fiji: A Short History*, Sydney: George Allen & Unwin.

Wilkinson, P. F. (1997) 'Jamaican tourism: from dependency theory to world-economy approach', in D. G. Lockhart and D. Drakakis-Smith (eds) *Island Tourism: Trends and Prospects*, London: Pinter, pp. 181–204.

Wilson, D. (1997) 'Paradoxes of tourism in Goa', *Annals of Tourism Research* 24: 52–75.

Woodcock, K. and France, L. (1994) 'Development theory applied to tourism in Jamaica', in A. V. Seaton (ed.) *Tourism: The State of the Art*, Chichester: Wiley, pp. 110–19.

Young, J. (1993) 'Sailing to Levuka: the cultural significance of the island schooners in the late 19th century', *Journal of Pacific History* 28, 36–52.

9 Neocolonialism, dependency and external control of Africa's tourism industry

A case study of wildlife safari tourism in Kenya

John S. Akama

Introduction

In most non-Western societies, particularly in Africa, that underwent colonial rule, external interest groups have over the years played a significant, if not a dominant role, in the development of tourism in those societies. In the case of Kenya, the development of tourism, particularly the development of wildlife safari tourism, is closely linked to the era of colonialism in the African continent. Indeed, it can be argued that the current forms of tourism development initiatives in Kenya and most other African countries are still, to a large extent, influenced by Western ideological values, and mainly respond to external economic interests. Thus the development of tourism in Kenya, as is the case with most developing countries, conforms to historical and economic structures of colonialism and external control.

Moreover, the nature of international tourism development as a 'luxury and pleasure seeking industry' usually entails, predominantly, rich tourists from the metropolis (mainly tourists from developed northern countries) visiting and coming to enjoy tourist attractions in the periphery (mainly poor and resource scarce countries in the South) (Britton 1982). Usually, these forms of international tourism development accentuate the economic structure of dependency on external market demand. These lead to 'alien' forms of tourism development (i.e. the establishment of enclave tourism resorts in African countries) to which local people cannot relate and respond to, either socially or economically (Williams 1993). In consequence the management and long-term sustenance of the tourism establishment, in most instances, depends on external control and support. This accentuates existing neocolonial tendencies and reinforcement of structures of economic dependency in developing countries, particularly in Africa.

Using the case of Kenya, the study traces the origins of wildlife safari tourism in the country from the colonial period to the present. The study identifies the underlying political and economic factors that have, over the years, influenced the development of wildlife safari tourism in Kenya as they relate to neocolonialism and economic dependency. Most of the information used in this research was acquired from government policy documents, national development plans, secondary literature, informal discussions and dialogue with stakeholders in the tourism industry and personal observations.

As the Kenyan case study indicates, the initial investment costs for large-scale, capital intensive tourism projects are usually too high for African governments and indigenous investors and, therefore, must depend on external capital investment mainly from multinational conglomerates. These structures of economic dependency usually lead to high leakages of the tourism revenues to external sources (Britton 1982; Oglethorpe 1984). In this regard, not much of the tourism revenues remain in developing countries to be utilised in various processes of socio-economic development. These forms of tourism development initiatives usually lead to situations of further marginalisation and socio-economic under-development.

Historical background: colonialism and the era of big-game hunting

The evolution of wildlife safari tourism in Kenya and most other African countries has its origins in the period of big-game hunting expeditions by pioneer European and North American adventurers and fortune seekers. From the mid-nineteenth century, an increasing number of Western adventure seekers, professional and amateur hunters started to venture into the hinterland of East Africa. The establishment of colonial rule over the East Africa Protectorate (the present Kenya) in 1895 and the subsequent development of centralised political and administrative institutions created the required initial socio-economic environment for the development of tourism activities.

The creation of colonial institutions of governance, for instance, engendered conditions of relative socio-political stability and the maintenance of law and order which encouraged pioneer Western travellers to venture into the East Africa hinterland. A major recreational activity undertaken by these pioneer Westerners was big-game safari hunting. In fact, the period between 1900 and 1945 in East Africa is generally referred to in popular literature as the 'Era of Big Game Hunting'.

During the initial period of European colonialism in Africa, and indeed in the rest of the Third World, the recreational phenomenon of big-game hunting was perceived as a major symbol of European dominance over nature in particular and society in general. In consequence, big-game hunting was a major determinant of class and socio-political power (Anderson 1987; Mackenzie 1987). Thus, most of the pioneer Westerners who undertook safari hunting expeditions in Africa were affluent travellers, high-ranking government officials, politicians and members of the aristocracy.

The famous pioneer travellers to East Africa and big-game safari hunters include such people as Theodore Roosevelt, John Muir, Frederick Lugard, Fredrick Jackson, Abel Chapman, William Baullie, Geoffrey Archer and Robert Coryndon (Anderson 1987). For instance in his most widely published safari to East Africa, which lasted between April 1909 and March 1910, the then US President Theodore Roosevelt travelled with over 200 trackers, skinners, porters and gun bearers. Roosevelt shot, preserved and shipped to Washington DC more than 3,000 specimens of African game.

Most of the pioneer safari hunters provided detailed accounts of their hunting exploits in the overseas colonies, whenever they returned to the West. Others wrote adventure books based on their big-game hunting exploits (Nash 1982; Mackenzie 1987). For instance, a British aristocrat and a professional hunter, Abel Chapman, wrote an adventure classic in 1908 entitled *On Safari,* where he recounts his spectacular hunting escapades in the East Africa savannas. He argues here that the big-game traveller-sportsman was the best customer of the East Africa colony and game was its best asset (Nash 1982). In the following year, 1909, an American big-game hunter, William Baullie, wrote another hunting classic titled *The Master of the Game*, with an introduction by Theodore Roosevelt. Part of the book's introduction reads, 'there were still a few remote places (on the face of the earth) where one had to hunt in order to eat and where settlers had to wage war against the game in the manner of the primitive man' (Nash 1982: 354). It should be stated that these safari hunting classics are still popular, and continue to reinforce Western perceptions and images of Africa in general, and Kenya in particular, as a wildlife 'Eden'.

It should also be noted that in most instances, during the initial stage of development of tourism in Kenya, as is the case with the other African countries, there was minimal interaction between Western travellers and indigenous African people. Similar characteristics of tourism development have been found in other Third World destinations (Harrison 1995; Douglas 1997). Perhaps the only form of interaction that existed between the class in power and the governed was a 'master-servant' relationship. Africans were mainly hired to work in servile positions as gardeners, porters, cleaners, waiters, cooks and guards.

Furthermore, initial development of tourism and hospitality facilities and infrastructure was mainly undertaken by external interest groups without much involvement of indigenous communities. For instance, the first conventional hotels and lodge facilities to be developed in Kenya were mainly built by resident European developers. These initial facilities include Hotel Stanley (the present New Stanley) in 1890, the Nairobi Club in 1891, the Norfolk Hotel in 1904 and the Commercial and Express Hotel in 1906. Most of these accommodation and hospitality facilities were built in Nairobi, which became the hub of commerce, business and administration in the East Africa region (Bosire 1995).

The creation of wildlife parks

There was wanton destruction of the African savanna wildlife by European amateur and professional hunters in search of prized trophies during the period of big-game hunting in East Africa. During this period of accelerated destruction of wildlife in the African colonies, pioneer Western conservationists realised that if the destruction was not checked, the end result would be extinction. Thus, the conservationists raised concern about excessive destruction of the savanna wildlife in the African continent. In consequence, the colonial government started to formulate and promulgate various laws aimed at the protection of Africa's unique wildlife attractions and the promotion of organised safari tourism activities in protected wildlife parks and reserves (Kenya Government 1957; Lusigi 1978).

In 1939, for instance, the British government, as a result of pressure from British conservationists, appointed a game committee to study and make recommendations regarding setting up game parks in Kenya and other colonies in Africa. The committee was mainly composed of British naturalists, aristocrats, explorers and top administrative officials. The committee was to plan the location, extension, constitution, control and management of game parks and; the forms of recreational activities that should be permitted in the parks. Accordingly, the game committee made certain recommendations that were approved by the colonial legislature in 1945. The recommendations of the game committee led to the creation of the pioneer national parks in Kenya which included Nairobi in 1966, Amboseli in 1947, Tsavo in 1948 and Mt Kenya in 1949. The committee recommended that for wildlife to be effectively protected from human impacts, the parks should be:

a) Under public control, the boundaries of which should not be altered or any portion be capable of alienation except by competent legislative authority.

b) Set aside for the propagation, protection and preservation of objects of aesthetic, geological, prehistoric, archaeological, or scientific interest for the benefit and advantage of the general public.

c) In which hunting, killing, or capturing of fauna and destruction or collection of flora is prohibited except by or under the direction of park authorities.

(Lusigi 1978)

Thus, the initiation of the pioneer wildlife conservation policies and tourism programmes were aimed at protecting wildlife from perceived destructive forces of humans. Wildlife conservationists and government officials felt that, for wildlife in the East Africa Protectorate to be adequately and effectively protected, nature conservation areas had to be established and boundaries demarcated which separated wildlife from development activities. Consequently, the pioneer state wildlife conservation policies and programmes in Kenya were aimed at protecting the savanna game from:

a) The skin hunters who seek and kill game solely for their skins, leaving carcasses to vultures.
b) Natives who cannot be made to understand the advantages of a closed season.
c) The wanton sportsmen who shoot females and who kill large numbers of males on the chance of securing a good specimen trophy.

(Ibid.)

In part, these forms of wildlife management policies and programmes were a consequence of conservation and administrative officials' Western experience and environmental values. Due to the rapid transformation of nature and the disappearance of most wildlife in the West, particularly during the industrial revolution, the general perception among pioneer naturalists was that most human land use practices were incompatible with the principles of nature conservation in general, and wildlife protection in particular.

It should be stated that, in most instances, the pioneer national parks were created without due consideration of the existing social and ecological processes in the places where the parks were located. The demarcation of the park boundaries did not take into consideration socio-economic factors as they relate to indigenous African communities. Moreover, the underlying concept among government officials and park management was that the indigenous resource use methods were destructive to wildlife, and that they were also incompatible with the development of wildlife safari tourism activities. Officials were faced with different natural resources utilisation methods, such as traditional subsistence hunting, pastoralism and shifting cultivation, and they had difficulties in evaluating and understanding these resource use practices.

Most often government officials and conservationists classified African modes of natural resource use as at best 'unprogressive' and at worst 'barbaric' and to be eliminated. Local people were prohibited from entering the park and utilising the existing park resources including pasture, wildlife, water and fuelwood. Ironically, these were the very resources which indigenous African communities depended upon for their sustenance. Thus, whereas wildlife safari tourism, an entirely European recreational phenomenon, was allowed in the protected game parks, subsistence hunting by indigenous people was banned and was, officially, classified as 'poaching'. As will be shown below, to a large extent these forms of wildlife conservation principles and tourism development initiatives have persisted into the post-colonial period. Also, over the years, the national parks and reserves have developed into major centres of safari tourism activities, and an increasing number of international tourists, particularly from Western countries, visit the parks for wildlife viewing and photographing.

External control and postcolonial tourism

At independence in 1963, the Kenya government inherited a colonial economy that was characterised by inequitable distribution of resources, high rates of unemployment and poor living standards among the indigenous population. Furthermore, the country's economic activities were controlled mainly by expatriates, who had relatively high standards of living *vis-à-vis* those of the indigenous population. In the case of tourism (as shown above), the initial development of tourism and hospitality facilities in the country was mainly initiated by resident European developers and the colonial government. As a consequence, the initial development of tourism in Kenya was colonial in orientation and mainly served the social and economic interests of the expatriate community and international tourists.

With the realisation of the importance of tourism development in generating much sought after foreign exchange, the Kenya government turned to foreign and multinational investors to provide initial capital for the establishment and development of large-scale tourism and hospitality facilities. In this regard, the government adopted an 'open-door' *laissez-faire* policy towards multinational tourism investors and developers. Furthermore, the government introduced specific financial incentives such as tax concessions, favourable fiscal policies for external capital investment and profit repatriation. These financial incentives were aimed at attracting increased and accelerated foreign capital investment in tourism and hospitality.

Due to increased foreign investment in Kenya's tourism and hospitality industry over the years, there is increased ownership and management of the industry by foreign and multinational companies. Some of the international tourism companies that have invested in Kenya's tourism and hospitality industry include Hayes & Jarvis, Lonrho Corporation, United Tours Companies, Kuoni, Africa Club, Universal Safari Tours, Pollman's, France Russo and Grand Viaggi. These multinational tourism companies have established first-class hotel and lodge facilities, particularly in Mombasa, Nairobi, Malindi and in the country's popular national parks and reserves. Furthermore, it has been estimated that over 60 per cent of Kenya's tourism and hospitality establishment in major tourist centres is under foreign ownership and management (Sinclair 1990; Sindiga 2000).

Moreover, the state tourism policy has over the years mainly promoted the development of large-scale tourism projects such as beach resorts, high-rise hotels, lodges and restaurants. These forms of capital intense programmes tend to preclude local participation in tourism project design and management and local use of tourism resources. As Sindiga reported:

> The ground operation of the country's tourism industry reflects [this] outward-orientation. Typically a tour operator sends a micro-bus to the airport to collect tourists. Such visitors may be in an inclusive package

tour already paid for overseas. The tour firms, for example, Abercrombie and Kent, United Tour Company, Kuoni Worldwide, Thomas Cook, and Hayes and Jarvis, would likely be foreign owned, or a subsidiary of a foreign company. The firms take the tourists to an assigned hotel in Nairobi or Mombasa for overnight stay. On the following day, the tour operators take the tourists to a wildlife safari in one of the national parks. This safari lasts several days. The average length of stay for departing tourists in 1992 was 13.4 days. . . . At the end of the tour, the process is re-enacted in preparation for departure from the country.

(1996: 29)

The creation of tourism image

It can be stated that the nineteenth- and early twentieth-century image of Kenya and other African countries as a wilderness 'Eden' still persists and is the main factor attracting Western conservationists and tourists to Kenya (see Wels this volume). Furthermore, Western naturalists and scientists still play a significant role in the conservation of Kenya's wildlife. A number of Western conservation and tourism organisations have established offices in the country, and they act as watchdogs and assist the government in nature conservation and the promotion of wildlife safari tourism. The organisations include International Union for the Conservation of Nature (IUCN), World Wide Fund for Nature (WWF), African Wildlife Foundation (AWF), the Max Planck Institute and Frankfurt Zoo. Indeed, such apparently local organisations as the East African Wildlife Society are dominated by Western membership while Kenya's tourism industry is deeply influenced by Western concerns and environmental groups (Akama *et al.* 1996). These tourism and conservation organisations recognise the remaining high concentration of tropical savanna game in Kenya as 'world heritage' that should not be allowed to disappear but should be protected for future generations.

It has been argued that to most Western middle-class people, pristine wilderness areas present alternatives for escape from what is perceived as harsh reality and stresses associated with urban life and industrial capitalism (Krippendorf 1987). Existing literature on tourist behaviour has shown that visitors are usually influenced by 'push and pull' factors when they make an initial decision to travel to far-off destinations (Mathieson and Wall 1982; Shaw and Williams 1994; Krippendorf 1987). The push factors include the urge to escape from the pressures of the workplace and the stresses associated with urban life and industrial capitalism whereas the pull factors include the urge to travel to different places in search of novelty and adventure. Thus, an overriding factor that makes tourists travel to far-off destinations is the demand to sightsee and experience different and exotic environments. However, this urge is contradictory in nature because of the extent to which both modernisation and tourism separates out objects from the societies and places that produced them (Krippendorf 1987).

Thus in the recent past, an increasing number of tourists from Europe and North America visit Kenya to view the country's unique savanna wildlife heritage. Most of the international tourists who visited Kenya in the 1990s, for instance, came from Western countries, mainly Germany, the United Kingdom, the United States, Italy and France with a significant portion visiting the country's national parks and reserves. Nature-based tourism also has become a lucrative business and tourism is currently one of Kenya's leading foreign exchange earners.

The Western perception of Kenya based on the nineteenth-century colonial exposition and safari adventures of the 1930s still persists and is being reinforced by tour companies and travel agencies' advertisements and the marketing of Kenya's nature attractions. Increasingly, more Western tourists are willing to pay for inclusive tour packages in order to visit what is perceived as the remaining 'wilderness Eden' in Kenya and other developing countries. Furthermore, the promotion and marketing of Kenya's tourism attractions in tourist generating countries is mainly done by overseas tour operators and travel agents. In this regard, it has been noted that tour companies play an important role in influencing tourist attitude, behaviour and preferences, and in determining the types and volume of tourists who visit a given tourist destination (Mathieson and Wall 1982; Shaw and Williams 1994).

Driven by the profit motive, most tour operators focus on marketing those tourist attractions that can yield immediate and maximum profit returns. Tour operators and promoters present partial information and images of Kenya's tourist attractions, as is the case with most other tourist destinations in developing countries. For instance, most tourist advertisements for Kenyan attractions in Western media mainly focus on the 'Big Five' (elephant, lion, rhino, cheetah and giraffe). In this regard, little effort is expended in giving a complete and accurate picture of Kenya's diverse nature attractions and other forms of tourist attractions (Sinclair 1990; Kibara 1994). Worse still, overseas tour operators have been accused of helping to reinforce existing stereotypes and images of Kenya, in particular, and Africa, in general. This is done in order to promote and increase the volume of tourist sales (Shaw and Williams 1994). Images of wild and darkest Africa, complete with roaring lions, trumpeting elephants, semi-naked and bare-breasted natives, are used to lure Westerners keen for exoticism and adventure.

The design and development of promotional messages and images that are used in sales promotion and the marketing of tour packages in tourist generating countries derive from and are usually based on existing dominant Western cultural values and economic systems. Further, the promotion and marketing of Third World tourist destinations in major generating countries in the West also derive from the forms of historical and economic relationships that exist between the developed and the less developed countries. As Morgan and Pritchard argue:

Tourism image (as constructed by tour operators and other tourism marketers) reveals as much about the power relations underpinning its construction, as it does about the specific tourism product or country it promotes. The images projected in brochures, billboards and television reveal the relationships between countries, between genders and between races and cultures. They are powerful images which reinforce particular ways of seeing the world and can restrict and channel people, countries, genders and sexes into certain mind-sets.

(1998: 6)

For instance, when marketing Kenya's attractions in tourist generating countries, the Maasai are usually presented as if they are the only African community that exists in Kenya (this notwithstanding the fact that Kenya is made up of more than 40 ethnic communities with diverse cultures and historical experiences). Thus, when tourists visit Kenya for a wildlife safari, they are also supposed to catch a glimpse of the exotic Africa culture as represented by the Maasai tribesmen. Consequently, in tourism circles, wildlife and the Maasai are usually wrapped together as one and the same thing. The African culture that the international tourists are presented with is that of the Maasai tribesmen and their physical adornments, dance and other Maasai cultural artefacts. The Kenyan tourism image is constructed and reconstructed to revolve around wildlife and the Maasai image and thus, the tourist image of the Maasai does not appear to have changed since early European explorers and adventure seekers first encountered the Maasai over 200 years ago.

So when Kenya's tourist attractions are marketed, the Maasai are prominently featured in brochures, advertisements, electronic media and other forms of tourism commercials that promote Kenya as a leading tourism 'mecca' in Africa. Scenes of the Maasai dressed in red ochre shuka and/or traditional regalia are juxtaposed with the 'Big Five' and are promoted as ideal African tourist attractions. The Maasai Moran (youthful warriors), carrying traditional long spears and clubs, are projected in the media as people who 'walk-tall' amidst the deadly Africa wildlife. Scenes of Maasai livestock are also projected in commercials, grazing in harmony with other savanna herbivores such as antelopes, zebra, wildebeest, buffalo and elephants.

The tourism images of harmonious co-existence between the Maasai and the savanna wildlife may have been tenable in the period preceding the creation of state protected game parks and the establishment of tourism facilities and infrastructure on land previously owned by the Maasai. In reality, the Maasai are often in severe and persistent conflict with park wildlife over grazing and water resources (the wildlife parks were created in important dry season grazing ranges). As discussed below, the situation has been accentuated by state tourism and wildlife conservation policies that focus narrowly on the protection of park wildlife for foreign tourists without any involvement of the local people in the management and utilisation of these resources.

Exclusion of local people from tourism

Kenya's tourism activities are spatially constrained to a few locations in the popular wildlife parks. The majority of Kenyan people in most regions of the country do not receive any form of direct monetary benefit from the industry. Furthermore, few people who live at or near tourist attractions and facilities receive jobs, even relatively lowly ones, in local tourism and hospitality establishments. Also, due to the increasing trend of inclusive tour packages using only a limited number of destinations on the few popular wildlife parks, fewer tourist receipts are reaching Kenyans at a grassroots level (Bachmann 1988; Sinclair 1990; Akama 1996). Consequently, the local people, who bear most of the costs of tourism development and wildlife conservation, do not receive any form of direct monetary benefits from the tourism industry. Some of the tourism costs incurred by the local people include water pollution, as raw sewage from the tourist hotels and lodges drains directly into the local water systems, and the disruption of indigenous cultures by mass tourism activities (Bachmann 1988; Sinclair 1990; Kibara 1994). Also in certain locations prime agricultural land, which could have been otherwise used for local food production and livestock rearing, is used for tourism development and wildlife preservation.

Thus, while the local people bear the costs of tourism development and wildlife conservation they in return receive insignificant direct monetary benefits. It has been estimated that only between 2 per cent and 5 per cent of Kenya's total tourism receipts trickle down to the populace at the grassroots level, in forms of low paying and servile jobs, and the selling of souvenirs and agricultural produce (Bachmann 1988; Sinclair 1990). As a consequence, while the tourism industry achieves considerable profit, few financial resources are allocated for local development. Tourism and wildlife conservation benefits to households or community are uncertain and are possibly non-existent.

The most extreme example of shifting the cost of wildlife conservation and safari tourism development is the fact that cultivators and pastoralists are not allowed to protect themselves or their property from wildlife despite considerable injury and severe damage to farms and livestock (Kenya Wildlife Service 1990; Akama 1996). State law prohibits any form of destruction and killing of wildlife. Consequently, peasants are reduced to guarding crops and livestock by making noise, beating drums and making night fires so that someone else may make profit from tourists willing to view and photograph an animal local opinion would wish dead. Hence, local people's attitudes towards wildlife protected areas varies from that of indifference to intense hostility (Lusigi 1978; Akama 1996).

Moreover, the concept of setting aside nature areas as protected parks may at best be inconceivable and at worst repulsive to rural African cultures (close to 80 per cent of Kenya's population reside in rural areas). Also, the wildlife parks still conjure images of the harsh colonial legacy of wildlife preservation and the establishment of protected areas (Akama *et al.* 1996). The colonial conservation policies and laws made traditional subsistence hunting an illegal and punishable offence. Many Africans were imprisoned on poaching related offences. In certain instances, whole communities, such as the Waliangulu of Southern Kenya, ended up in prison, which is similar to what happened to the Ik of Northern Uganda for much the same reason – virtually every adult in these communities was a subsistence hunter.

In recent years, the Kenya wildlife service and tourism groups have attempted to initiate community based wildlife conservation and tourism development in areas adjacent to wildlife parks and reserves. However, it should be stated that most of the so-called community based programmes have ended up with the co-opting of local elites in wildlife conservation and tourism development and with little meaningful involvement of the majority of rural peasants, particularly in project design and management. For instance, since the mid-1980s, the Kenya Wildlife Service has been implementing community-based wildlife tourism projects in areas around Amboseli National Park and Maasai Mara National Reserve (Sindiga 2000). The new policy is aimed at encouraging the local people to form wildlife conservation associations to participate directly in wildlife safari tourism development. However, these wildlife associations have ended up being dominated by local elites who monopolise and control most of the tourism revenues accruing from camping and lodge concessions, and gate entrance levies.

Conclusion

As this chapter demonstrates, the evolution of wildlife safari tourism in Kenya has its origins in the colonial period when pioneer Western adventure seekers ventured into the hinterland of the African continent to undertake in big-game hunting activities. Furthermore, the initial development of tourism and hospitality facilities was, mainly, undertaken by resident European settlers and the colonial government with minimal involvement of local African communities. The study argues that these trends in tourism development have persisted to the present time. Over the years, the postcolonial Kenyan government has encouraged an 'open door' policy towards external private and multinational tourism investors and developers.

The government has initiated policy initiatives that create a conducive socio-economic environment to attract external and multinational investments in the tourism sector. The policy initiatives include tax concessions, and the creation of favourable fiscal regulations for capital investment and profit repatriation. These policy initiatives have succeeded in attracting increased multinational capital investment in Kenya's tourism industry. As a consequence, it has been estimated that over 60 per cent of Kenya's tourism establishment is under foreign ownership and management. Furthermore government policy initiatives which promote the development of large-scale, capital intensive tourism projects tend to preclude indigenous and local investment in the tourism sector. Also, as the study shows, the country's wildlife based tourism programmes tend to preclude local people in the management and utilisation of the local wildlife and tourism resources. Thus, while the local people bear most of the direct and indirect costs of wildlife conservation and tourism development, they in return receive minimal and insignificant benefits from the tourism sector.

References

Akama, J. S. (1996) *Wildlife Conservation in Kenya: A Political-Ecological Analysis of Nairobi and Tsavo Regions*, Washington DC: African Development Foundation.

Akama, J. S., Lant, C. and Burnett, W. (1996) 'A political-ecology approach to wildlife conservation in Kenya', *Environmental Values* 5, 4, 335–47.

Anderson, D. (1987) 'The scramble for Eden: Past, present and future in Africa', in D. Anderson and R. Grove (eds) *Conservation in Africa: People, Policies and Practice*, New York: Cambridge University Press, pp. 1–12.

Bachmann, P. (1988) *Tourism in Kenya: A Basic Need for Whom?*, Berne: Peter Lang.

Bosire, S.M. (1995) *Training Hospitality Managers in Kenya: A Re-appraisal*, unpublished staff seminar, Department of Tourism, Moi University, Eldoret, Kenya.

Britton, S. (1982) 'The political economy of tourism in the Third World', *Annals of Research* 9, 331–58.

Douglas, N. (1997) 'Applying the life cycle model to Melanesia', *Annals of Tourism Research* 24, 1–22.

Harrison, D. (1995) 'Development of tourism in Swaziland', *Annals of Tourism Research* 22, 1, 135–54.

Kenya Government (1957) *Annual Report for 1956–1957*, Nairobi, Kenya.

Kenya Wildlife Service (1990) *A Policy Framework and Development Programmes, 1991–1996*, Kenya Wildlife Service, Nairobi.

Kibara, D. (1994) 'Tourism Development in kenya: the government's involvement and influence', unpublished Master's dissertation, University of Surrey.

Krippendorf, J. (1987) *The Holiday Makers*, London: Heinemann.

Lusigi, W. J. (1978) *Planning Human Activities on Protected Natural Ecosystem*, Germany: Verlag.

Mackenzie, J. M. (1987) 'Chivalry, social darwinism and ritualized killing: the hunting ethos in Central Africa up to 1914', in D. Anderson and R. Grove (eds) *Conservation in Africa: People, Policies and Practice*, New York: Cambridge University Press, pp. 41–62.

Mathieson, A. and Walls, G. (1982) *Tourism: Economic, Physical and Social Impacts*, London: Longman.

Morgan, N. and Pritchard, A. (1998) *Tourism Promotion and Power. Creating Images. Creating Identities*, New York: John Wiley & Sons.

Nash, R. (1982) *Wilderness and the American Mind*, London: Yale University Press.
Oglethorpe, R. (1984) 'Tourism in Malta: a crisis of dependence', *Leisure Studies* 3, 2, 147–61.
Shaw, G. and Williams A. (1994) *Critical Issues in Tourism: A Geographical Perspective*, Oxford: Blackwell.
Sinclair, M. T. (1990) *Tourism Development in Kenya*, Washington, DC: World Bank.
Sindiga, I. (1996) 'Domestic Tourism in Kenya', *Annals of Tourism Research* 23, 1, 19–31.
Sindiga, I (2000) *Tourism and African Development: Change and Challenge of Tourism in Kenya*, Aldershot: Ashgate Publishing.
Williams, M. (1993) 'An expansion of the tourism life site cycle model: The case of Minorca (Spain)', *The Journal of Tourism Studies* 4, 2, 24–32.

10 Postcolonial conflict inherent in the involvement of cultural tourism in creating new national myths in Hong Kong

Hilary du Cros

Introduction

Hong Kong has survived the challenge of the 1997 Handover from British to Mainland Chinese control and the economic reversal caused by the unrelated events of the Asian financial downturn and the September 11 disaster. Although the latter has had more impact on the lives of everyday people, the Handover was a cause for a certain amount of nervousness about the future with concerns raised about change to civil, administrative and economic freedoms. The passing of colonialism has allowed some examination of Hong Kong cultural identity in government programmes and through other venues, but much of the focus has been on maintaining its economic advantage in the region as the economy has slowed in growth. Moreover, the tourism market appeal of Hong Kong as a brand changed after the Handover, causing some disappointment among economic forecasters and tourism authorities. A number of aggressive new marketing strategies have been adopted, which present an artificial summary of its cultural identity, race relations and attitudes to colonialism. One such strategy is the heavy emphasis on it being a marriage of East and West characteristics that ignores the ambivalence in the relationship past and present. Yet even more strategies are planned, some of which may not be sensitive to postcolonial changes in the Hong Kong Chinese cultural identity in particular, and which may also fail with the tourists the authorities are trying to target.

Hong Kong appears to be promulgating a vision of itself as a progressive and innovative predominantly Chinese cosmopolitan society (Culture and Heritage Commission 2002), and is using this myth to empower its population to feel more secure about their Chinese cultural identity in relation to that of Mainland China and other centres of Chinese culture. This mission can cause conflict when it collides with recent efforts to position Hong Kong as a tourism destination that uses its 'Western-ness' and sometimes its colonial heritage as a distancing factor from other Asian-Chinese cities. Hong Kong is often also seen as 'Safe Asia', an easy destination for Western tourists to absorb, and possibly an easy option also for the Mainland tourists to consume Western culture as it has elements to satisfy both (du Cros 2002a).

Hong Kong is used here as a central case study with which to explore such conflicts over the symbolic nature of some of its heritage attractions in a postcolonial context. Some comparisons are made, however, with other postcolonial destinations, namely Macau and Singapore. The common theme centres on negotiating notions of cultural integration and colonialism in the marketing of sustainable heritage tourism in postcolonial places in south-east Asia.

National myths and loaded symbols

It could be argued that Hong Kong cannot be either postcolonial or able to establish a new national myth on its own, as it is now part of China. As part of China it is no longer a British colony but a type of Chinese colony or province, and accordingly its vision of itself should not differ from that which the Mainland holds for itself. But Hong Kong has a uniquely Chinese-British history and has developed in a very different way, so that 'national myth' in a more unitary or abstract sense applies without disenfranchising China's new sovereignty over the territory.

Cultural identity can be defined as 'a snapshot of unfolding meanings relating to self-nomination or ascription by others . . . it relates to nodal points in cultural meaning, most notably class, gender, race, ethnicity, nation and age' (Barker 2000). National identity is therefore built on cultural identity for a nation-state and expressed through symbols and discourses so that nations are not only political systems but also systems of cultural representation. These symbols can be evoked in national myths that act as a symbolic guide or map of meaning and significance for a society. If they are mixed with loaded symbols, that is, ones that can evoke a passionate response from members of the society as it clashes with some aspect of their cultural identity, then the myth is likely to be less convincing in its role as guide. In a more cynical way, Lowenthal (1996: 129) notes that, 'out of some legendary kernel of truth, each corporate group harvests a crop of delusory faiths . . . that sustains (it)'. However, truth can be relative, and having someone else's truth forced upon you as either a tourist or host can be unpopular. Hence, care should be taken with the marketing of heritage attractions regarding which historical narrative is being evoked by its subliminal or not so subliminal product associations.

The types of attractions and potential attractions that are and could become part of Hong Kong's postcolonial national myth making process include museums, heritage places, heritage trails, public statuary, remnant examples of colonial streetlife (such as rickshaws) and Hong Kong Disneyland (still under construction). These examples all share one common aspect and that is that they have been or will be used as part of Hong Kong's promotion of itself to mainly Western tourists (some of which come from the former colonial country itself) as a unique international and cultural destination in south-east Asia. Moreover, this promotion and product creation is occurring in a postcolonial context with all the possible ambivalence that that implies. Hence, there are at least three main issues:

1   The time/distance from unpleasant associations of colonialism and how willing hosts are to allow the integration or decommissioning of elements of the colonial past as part of tourism product promotion;
2   the attitude of authorities past and present to local community concerns, cultural identity and new national myths; and
3   the nature of loaded symbols that could create trouble for newly forged postcolonial cultural identities and national myths, e.g. public statuary, pith helmets, rickshaws and theme parks with a strong cultural message.

The treatment of culture and heritage in Hong Kong

The colonial period sets the scene for current attitudes to culture and heritage and so a brief discussion of cultural identity, race relations and notions of empire is merited for Hong Kong. The end of British colonial rule came at midnight on 30 June 1997. It was decided that while Hong Kong should continue to serve China's economic and political strategies it should also retain a large degree of autonomy. This autonomy would be guaranteed by the 'one country, two systems' policy under which Hong Kong could retain its legal system, freedom of speech, land rights and way of life while recognising China's sovereignty. On 1 July Hong Kong became a Special Administrative Region (SAR) of the People's Republic of China, to be administered by Hong Kong people using major elements of the previous colonial system for the next 50 years under the newly established Basic Law (Hong Kong's jointly agreed declaration of its status as a SAR of China).

The colonial derived legislation associated with the protection of heritage assets is actually very weak, and responsibility for their conservation and presentation for tourism is fragmented. This poor legislative protection occurs mainly as a result of the urban planning model adopted by the colonial government that has placed such a high value on real estate development as a source of government revenue through property taxes. Consequently, efforts to review the suite of laws that affect urban planning, renewal and heritage conservation continue to be mired in bureaucratic inertia. It seems that only the media can invigorate uproar about the proposed demolition of important buildings (see the recent Kom Tong Hall debate in the *South China Morning Post* 2002c, 2002d, 2002f). Often, the acquisition of such heritage places by the government is the only way to save them. That there is a growing concern about such demolitions in district councils and local communities is a relatively new development. It suggests there is growing distress about landmarks and settings that would otherwise symbolise Hong Kong's rich cultural and historical development being removed for development (Abbas 1997; Heng 2002). Such places would provide an anchor for the Hong Kong Chinese community in the rapidly changing and globalising world in a way that would reassure those who are committed to a long-term residence in Hong Kong.

In understanding the meaning of cultural identity for Hong Kong Chinese, elements of transience have always had a role. Many Chinese living here are either first or second generation migrants and some people were so concerned by events such as Tiananmen Square and the Handover that they have residences overseas and have sought additional pass-ports. The care and protection of heritage assets that are not portable is not often seen as vital by transient groups in industrialised societies, hence the long-term commitment to Hong Kong that appears to be emerging post-Handover is a major factor in agitating for the conservation of Chinese cultural heritage particularly. It can also be seen in the two large new museums, the History Museum in Kowloon and the Heritage Museum in Shatin, which have a strong local (and sometimes overlapping) heritage focus.

Cultural identity is also important to the recently appointed Culture and Heritage Commission which advises the Hong Kong SAR Government on overall policies and funding priorities in the development and promotion of culture and heritage. The Commission aims at: enhancing the quality of life of Hong Kong people; fostering a sense of belonging and cultural identity among the public; and developing Hong Kong into a centre of international cultural exchange (Culture and Heritage Commission 2002).

The Culture and Heritage Commission published a consultation paper entitled *Gathering of Talents for Continual Innovation* in mid-March 2001 and has received favourable public support for its proposed principles and strategies for promoting cultural development in Hong Kong. Public consultation on the new cultural policy has revealed that beside the support, 'some submissions even convey a sense of longing for a bright future in the development of culture in Hong Kong' (Home Affairs Department (HAD) 2001). The use of the word 'longing' is rather telling and is likely to refer in part to those submissions that reveal a frustration regarding Hong Kong's poor record on heritage protection and promotion of locally based artistic talent.

The Hong Kong government has also initiated more cultural programmes for schools and the local community with an emphasis on learning about uniquely Hong Kong and Chinese traditions such as Cantonese Opera, traditional dancing, music and handicrafts (Hong Kong Leisure and Cultural Services Department (LCSD) 2002). In creating such programmes, the role of other cultures in the development of Hong Kong is rarely presented. After being the premier focus of school history lessons for many years, the British period of occupation has been reduced in importance in order to highlight Hong Kong's Chinese heritage and its close links with the Mainland.

The first exhibition to stress the Chinese heritage strongly was the Antiquities and Monuments Office display for the public at its offices in Nathan Road in 1998, just after the Handover. It comprised archaeological artefacts from the Ma Wan site with a sequence dating back to the Middle Neolithic nearly 6,000 years ago. This exhibition hailed the discovery as being highly significant because: it showed there was 'a profound tie with the Mainland' early on; it was a 'successful joint excavation with the Mainland archaeologists'; the 'Mainland experts recognised Hong Kong's achievement'; and it 'heightened public awareness of local history' (Antiquities and Monuments Office (AMO) 1998).

The exhibition was followed by the production of a CD Rom and education kits for schools on heritage, with pamphlets and posters all referring to the continuous sequence of 6,000 years of Hong Kong history. This information has been produced in parallel with much of the marketing by the Hong Kong Tourism Board (HKTB) and is occasionally found by tourists in museums. Most Hong Kong museums, however, have adopted its underlying message of long occupation prior to colonial rule as a kind of historical revisionism. This revisionism and its implications for a strengthening partnership with the Mainland are emphasised far more in their information than that of HKTB. However, the HKTB has a tendency to downplay the Handover and its implications for a closer relationship with China in the interests of maintaining its traditional Western and Japanese markets that are apprehensive about Chinese communism.

Visits to museums may sometimes leave tourists somewhat disappointed after the build-up to or glossing over of East/West cultural relationships that the Hong Kong Tourism Board has promoted in its campaigns. The Eastern and Western elements of the colonial history of Hong Kong are not as easy for tourists to experience in such displays as current marketing suggests. One other possible reason for this is that, historically, there has been a sequence of official discrimination (before 1945), and unofficial discrimination since then, which has led to a very superficial fusion of both elements. Post-Handover this may have led to a backlash against presenting much about the colonial period among museum curators as they view it with a degree of ambivalence based on their own experiences. It can also be seen in the attitude about practising the English language among many in the Chinese Hong Kong community, which is the despair of those wanting to promote the SAR as an international multicultural community to business and tourism interests. This issue and others, such as continuing self-segregation of the Expatriates/Anglophones and the Chinese communities, tend to indicate that East and West characteristics are present in Hong Kong, but not as integrated as the more multicultural Singapore. Singapore has three main cultural groups that are more equal in size and English is often used as a common language among them. Singapore has also been independent of British influence longer and this is evident in the way some museum exhibitions discuss the colonial past.

Britain should have its role in this ambivalence recognised, as much of Hong Kong could be considered shared or 'mutual heritage' under the current International Council on Monuments and Sites (ICOMOS) definition (ICOMOS 2004). What was the attitude to heritage and British colonial imperialism just prior to and after the Handover? How has and will it affect tourism and the development of local Chinese cultural identity and national myth? What symbols, loaded and otherwise, has the colonial period left that will influence that myth for Hong Kong?

Jacobs (1996) comments that 'empire' is still a living concept in the English mentality and is not just manifested in the sentimental retention of nineteenth- and early twentieth-century buildings in Hong Kong or London. It can be seen as an influence on the construction of new building schemes that are acting as 'place-events' and as symbols of an architectural revitalisation of notions of England's global significance. One example is the redevelopment of Bank Junction, London. A triangular group of nineteenth-century commercial buildings was demolished during a 1990s redevelopment amid much protest from heritage groups. The development was knocked back by lower level planning authorities, but finally gained approval at the highest level with the Parliamentary Law Lords. Jacobs notes that they were trying to 'activate a memory of empire even while they display a seeming disregard for its built environment legacy' (Jacobs, 1996: 40). The urban renewal of this part of London was one such reaction to loss of empire, and the constant redevelopment of the Central district of Hong Kong towards taller and more spectacular buildings is probably another. Very little now remains of the commercial district and docks that symbolised the high colonial phase of Hong Kong prior to the Second World War.

The colonial built legacy that England has left Hong Kong not only includes the planning conditions that encouraged this precinct of multi-storey buildings but also an elaborate infrastructure of roads, services, housing (the New Towns) and the new Chek Lap Kok airport that it expected to become the hub of Asian air transport. Very little funding and thought was allocated to the conservation of colonial heritage buildings, unlike in Macau. The Macau colonial government made a special effort in the years prior to Macau's Handover to the People's Republic of China in 1999 to complete restorations of colonial heritage buildings and gardens, such as the restoration of buildings in the Sao Paulo district and those on the Avenida da Praia, near Taipa Village. These works provide tourists and the local community with a strong statement of the Portuguese historic presence as a colonial power. Recently, cultural heritage analyst Carla

Figueiredo noted that 'the Chinese government hasn't diminished any of the efforts' (*South China Morning Post* 2002e). It is an interesting question as to which strategy, Britain's or Portugal's, will turn out to be the longest lasting statement of power over a former colony and how these newly postcolonial societies will manage to integrate such legacies into local myths.

Marketing Hong Kong as a cultural tourism destination

The 'City of Life' brand came out of the 'Hong Kong is It' campaign devised in 1999 by the Hong Kong Tourism Association (later the Hong Kong Tourism Board). Awareness had struck that some of the destination's traditional markets might view Hong Kong differently as a result of the Handover. In fact, there had already been an immediate drop in Japanese tour groups after 1997. The English and some Europeans also seemed less willing to visit in the same numbers as before (Hong Kong Tourism Board 2002a).

At the time of writing, the latest from the Hong Kong Tourism Board is that 'Hong Kong is exhilarating':

> **City of Life: Hong Kong is it!** is a two-year celebration co-ordinated by the Hong Kong Tourism Board, Home Affairs Department, Leisure & Cultural Services Department, the Tourism Commission and Hong Kong's 18 District Councils, and sponsored by the Hong Kong Jockey Clubs Charities Trust.
>
> The programme showcases the events, festivals and activities that have helped to make Hong Kong the most popular destination in Asia. Both visitors and residents can explore Hong Kong's 18 Districts, which will highlight local events and attractions as well as international standard events.
>
> From now until March 2003, you are invited to participate in this celebration of our lifestyle, culture and traditions – a Living Fusion of East and West. You'll want to stay here for Life!
>
> (Hong Kong Tourism Board 2002b)

This also links into their 'cultural kaleidoscope' brochure series that features heritage attractions, including both tangible and intangible heritage assets. The Meet the People Itinerary section appears to be for Western tourists seeking tourism experiences of Chinese culture as it boasts English speaking experts who will help visitors 'delve deeper into Hong Kong's unique way of life'. It is reasonably comprehensive in its range of activities including classes in Tai Chi, Feng Shui, pearl and jade grading, antiques and Kung Fu. It also has information on architectural appreciation walks and promotes the main heritage and art museums again for that market. However, there is little or no information of a similar kind about

Western architecture and colonial history suitable for Chinese or other Asian tourists.

The success of this new strategy appears to be mixed. While providing high quality cultural tourism, it is unlikely to appeal to the masses. Its success will depend on how long such cultural tourists stay and how freely they spend. For mass tourism, however, tourism experiences, especially many cultural tourism experiences, have their basis in entertainment. To be successful, and therefore commercially viable, the tourism product must be manipulated and packaged in such a way so as to be easily consumed by the public (Cohen 1972). Clearly, learning opportunities can be created from the experiences, but their primary role is to entertain (Ritzer and Liska 1997). In Europe and North America, museums and art galleries that are developed to provide educational and cultural enlightenment have recognised that they are also in the entertainment business and have arranged their displays accordingly (Zeppel and Hall 1991; Tighe 1986; McDonald and Alsford 1989; Prideaux and Kininmont 1999). The reason is that only a small number of tourists really want to seek a deep learning experience when they travel. The rest are travelling for pleasure or escapist reasons and wish to participate in activities that will provide a sense of enjoyment.

Hence, HKTB's focus on Western tourists seeking deep experiences is supported by their ideal of attracting the cultural tourism market of mature, wealthy and reasonably discerning travellers from Europe, North America and Australia (McKercher *et al.* 2002). While this view is based on their experiences in the colonial period, it ignores the needs of other types of tourists and the changing appeal for them of postcolonial Hong Kong. Such change may not all be negative, as part of that postcolonial appeal to some younger Mainland Chinese tourists is the desire to come and experience Hong Kong as a prosperous, multicultural and freedom-loving society (du Cros 2002b).

Heritage tours

A number of private operators aside from HKTB also provide tours that visit heritage attractions either as incidental stopovers in between shopping or as specially packaged sightseeing trips. These tours are generally aimed at the Mainland Chinese market and do not usually include much aside from temples such as the Po Lin Monastery's Big Buddha and Wong Tai Sin Temple. Day tripping Hong Kong Chinese are also serviced by private tours rather than those organised by the HKTB, who are loath even to give them a pamphlet. So it appears that HKTB's mission regarding cultural tourism is to service the international Western tourists, not domestic or international Chinese tourists. This dichotomy may have also spread to other tourism bodies, such as the Tourism Commission, which is now responsible for tourism development strategies and policies. It is a new government body that has taken on this role that was HKTB's by default before its establishment.

Growing out of the mission outlined above, the Hong Kong Tourism Board promotes a range of natural and cultural tour products that include heritage attractions:

'Green City'
1 Bird-watching
2 Dolphin-watching
3 Guided Nature Walks: Tai Long Wan, Sai Kung – Enchanting Escape Hike
4 Guided Nature Walks: Dragon's Back – Coastal Vista
5 Guided Nature Walks: Lantau Island – Trails and Temples
6 Ocean Park Behind-the-Scenes Tours

'City by Heritage'
1 Heritage Tour
2 Echoes of Hong Kong Tour
3 Heritage and Architecture Walks: Hong Kong Island
4 Heritage and Architecture Walks: Kowloon
5 Heritage and Architecture Walks: New Territories

'City of Islands'
1 Hong Kong Back Garden Tour – Sai Kung
2 Islands Hopping Pass
3 Lantau Explorer Bus
4 Lantau Island Tour
5 Outlying Island Escapade
6 Tap Mun (Grass) Island Walk
7 The New Lantau Bus Company Day Pass

'City by Foot' (Do-it-yourself walks)
1 Back Streets of Central via the Mid-levels Escalator
2 Cheung Chau: Island Retreat
3 Lamma Island: Wild and Wonderful
4 Sha Tin Walk
5 Surprising Stanley
6 Tai Tam Reservoirs: In the City's Shadow
7 The Peak to Pokfulam: Picture Perfect
8 Yau Ma Tei and Mong Kok: Memorable Markets

(HKTB 2002b)

These are all tours designed to appeal to Western tourists and show that landscape and natural heritage are also becoming important in HKTB's tourism marketing strategy. More colonial and Western heritage assets are included, but not in a way that really deals with their role in colonial history clearly. The Second World War was a huge turning point in race relations and empire in Hong Kong, as the British were shown to be vulnerable, or as historian Jan Morris puts it, 'no longer could the British feel themselves in all ways superior to the Asiatics, and though the manners of racial prejudice were to linger on, its forms disappeared' (Morris 1997: 261). The Peak, for instance, was an example of such racial prejudice up until the Second World War, as it was a European enclave where Chinese were forbidden to build. Little of this history is noted in HKTB brochures or signage around The Peak. The location of colonial buildings and ruins looking down on the rest of the harbour and city made them past but still potent symbols of colonial power in Hong Kong and its too confronting history.

Heritage trails

Fragmentation of responsibility for managing heritage tourism products is also a problem for heritage trails. Heritage trails are used for tourism although this was not why they were established. They are not managed or presented for tourism by either the HKTB or the Tourism Commission. Instead, the Antiquities and Monument Office or local district councils are expected to provide the on-site management and interpretation. However, two official trails and one unofficial heritage trail are receiving some promotion or mention as part of the 18 districts strategy. This initiative has unwittingly put the lack of cooperation between stakeholders and active management for tourism into media focus at the Ping Shan Heritage Trail.

HKTB hoped to promote tourism attractions to Western tourists in each of the government districts to encourage some decentralisation of tourism spending. The district councils and the Hong Kong Tourism Board chose the attractions to promote and each month a different district took its turn. Problems arose when Yuen Long was made 'District of the Month' and the Ping Shan Heritage Trail became the 'signature attraction that visitors should not miss'. Unfortunately, this turned out to be not the case. The trail was the first and least successful of the Antiquities and Monuments Office's heritage trails aimed at locals and schools. It has had a chequered management history and many problems between stakeholders since opening in December 1993 (Cheung 1999). It seems likely that little of this was taken into account when the choice was made to promote it and very few of the stakeholders were consulted. Other than producing a new brochure for the trail, nothing new appears to have been done on-site or with the cooperation of the local residents to present it to tourists.

Accusations were made in the media that no one connected with this project had actually visited the site prior to increasing its profile. A newspaper reporter found that: signage was neglected making it difficult to follow the trail; key buildings along the trail were closed (some owners had withdrawn their support of the trail and there were other problems); most did not have brochures available; maintenance of some key structures was poor and the visual appeal of some structures was affected by garbage nearby. One lone Western tourist was seen wandering around lost (*South China Morning Post* 2002a).

The trail is also much shorter than the original (1993) one, probably because some of the heritage buildings in private hands have been closed to visitation since then after an unresolved dispute between the AMO and local clan leaders regarding visitor management and other issues (Cheung 1999). The more recent problems also show that the remaining trail is not being actively managed as a single unit by either the local clan or the AMO. An editorial in the *South China Morning Post* criticised the HKTB for not taking a more proactive role in the management of the trail before promoting it, saying 'even though the Board has no control over the trail's attractions – or most of Hong Kong's scenic spots for that matter – they should have at least made regular inspections of those they routinely promote to check that they remain scenic' (Editorial, *South China Morning Post* 2002a). Despite the poor visitor experience it provides and the negative attitude to tourism held by some members of the host community, the trail has not been removed from the generic promotion of heritage attractions of Hong Kong (Hong Kong Tourist Board 2002b) nor has its management improved.

Further conflicts over new marketing proposals

Further critiques have emerged of tourism authorities' insensitivity to the newly forming national myth of Hong Kong as a progressive, free, predominantly Chinese, yet somehow multicultural part of China. If these authorities had been more in tune with these trends they could have avoided more embarrassment in the media, about a proposal that was aired to establish a working rickshaw stand in Stanley, on the southern side of Hong Kong Island near the markets there. The rationale was that this would be 'more authentic' than the rickshaws that are set up for photo opportunities, and which do not pull round passengers, near the exit of the Star Ferry at Central. Rickshaws and sedan chairs were used in the past to transport the colonials, tourists and wealthy Chinese around Hong Kong mainly in the years before the Second World War. Before more formal public consultation on this issue started, a debate in the media occurred about how the revival of this old icon of colonial rule would sit with the current understanding of postcolonial cultural identity. Stacey Lo of the Economic Development and Labour Bureau simply stated that 'many people feel it is undesirable to see locals pulling foreigners'. The *South China Morning Post* commented more colourfully that many people 'are sure to find the image of pigeon-chested pensioners hauling bumper-sized tourists up and down Stanley's hills a tad disturbing' (*South China Morning Post* 2002b). Whether or not the locals to be hired were the same ones that touted rickshaws for photos near the Star Ferry, pulling Westerners in rickshaws was very much a potent symbol of empire and of the inequities of the old system of colonial rule at its worst. This would have been a clear case where such authenticity for Western tourists would have been too much for the local community and sections of the government that are developing a new national myth of Hong Kong as an integrated cosmopolitan society.

Decommissioning colonial symbols for successful use in tourism marketing

The problem encountered above with using such a contested symbol for tourism promotion and entertainment might not have been so difficult if time and distance had intervened and if the market targeted was not so obviously meant to include the population of a previous colonial power. It could not avoid being offensive given this context. It is likely that many of the tourists themselves would find it quite disturbing, unless perhaps they were Mainland Chinese who appear quite comfortable with the re-emergence of sedan chairs as a means of transport at scenic spots, such as Tiger Leaping Gorge in Yunnan. Chinese carrying or pulling Chinese (of similar sizes and ages) might not cause much of a comment in Hong Kong either. It is likely that the recent and more direct experience of the inequity of colonialism and the need to empower the Chinese after many years of official and unofficial discrimination is at the fore in the rickshaw debate in Hong Kong.

Singapore, as an older postcolonial society, has had time to come to terms with the more offensive aspects of its past to the point that it has developed a hazy nostalgia around some symbols. The Raffles Hotel, where Chinese were not allowed to drink in colonial times, now allows access to any who can afford it. Its façade, and the invitation to stop and have a 'Singapore Sling' cocktail, feature in most destination marketing. What is less well known is that the proposed sale of pith helmets to tourists in the hotel themed shop initially caused some comment among locals, although it seems unremarkable to most now.

Despite being an even newer postcolonial society than Hong Kong, Macau appears to have a less ambivalent attitude to colonial remains. Maybe this is because Macau's principal industries are both recreational, namely gambling in casinos and tourism, so it has less of a mission to prove itself as a Chinese international city. Tourism marketing treats its Portuguese and Chinese heritage assets as equally important, while it received huge amounts of funding in the ten years prior to the Handover to conserve the former. One example of Macau's tolerance of or playful attitude towards its Portuguese heritage is with some of its public statuary, in particular, one statue near Taipa Village commemorating a local Portuguese poet in a small park, which is just incredibly effete with no seriously colonial message implied.

Hong Kong's official statuary was always of a more strait-laced Victorian nature with Queen Victoria herself represented in Statue Square until the Second World War. Apparently, the square was so named because of the large number of public statues of heads of government and important figures. However, during the Second World War the Japanese shipped almost all of them to Japan for scrap metal. When the war ended, Queen Victoria's statue was returned as it was recognised and rescued from a scrap heap by one of the occupying forces. It now sits in Victoria Park. The square still has one remaining statue, that of Sir Thomas Jackson, but no new figures have been added since the war or the Handover. The colonial authorities must have lost interest in such displays of 'empire' after their power was challenged by the Japanese and chose to concentrate on the economy and extensive building programmes instead. It also appears that the local community now has more affection for the two British style bronze lions cast in Shanghai in 1935 that flank the entrance to the new Hong Kong and Shanghai Bank across the road than it ever did for such statues. The paws are kept from tarnishing by the countless people who have stroked them for better fortune (Lim 2002). All these attractions and the square itself are covered by English language guidebooks and mentioned briefly in the heritage walks for Central in ways that are clearer and easier to locate than those elements of the Ping Shan Heritage Trail.

New purpose-built East/West fusion attractions

In contrast to the British experience, American cultural colonialism has yet to have a major impact on Hong Kong and maybe it never really will. It is interesting to follow this issue in relation to the politics of inclusion in the context of the design process for Hong Kong's Disneyland. The success of any cultural tourism attraction is reliant on it having appeal for the domestic tourists who are more likely to visit repeatedly. If those tourists have certain beliefs that could be considered cultural sensitive issues it should have an impact on how a landscape is considered. Even with a completely artificial American themed cultural landscape, such as the Disneyland being built in Hong Kong to open in 2005, Disney has had to carefully include the geomantic principles of Feng Shui in its design. Why? Because Disney knows it is in the best interests of the success of development, particularly after the lessons it learned in France, to make sure the cultural sensitivities of domestic tourists and local staff are not ignored (Ap 2000). It therefore wants local Chinese to feel comfortable visiting the place and for local staff to feel that it is a healthy and prosperous place to work.

At the time of writing this chapter, Disney is trying to incorporate the principles of Feng Shui in the design of the theme park without adding any overt Chinese attractions, such as a Chinese history themed activity area (Hong Kong Disneyland 2002). Domestic Chinese tourists as well as Mainland and other Asian tourists are the main market for this development (Ap 2000).

With the decommissioning of British colonialism after 1997, all areas of Hong Kong society appear to be showing a growing reliance on Feng Shui as part of their daily lives. For instance, very few buildings are built without some attention to Feng Shui principles even though many of these principles had their origins in servicing the functional needs of agricultural life (Hase and Sinn 1995). The positioning of rice fields, forests and village houses is very different to that required for multi-storey buildings, but nevertheless urban Chinese have adapted it to a postindustrial lifestyle with great fervour in many cases. One day it may even be integrated into town planning and housing construction legislation, as Feng Shui is seen as important to Hong Kong Chinese identity and ignoring it is increasingly being viewed as being un-Chinese. Hong Kong Disneyland will still be a strongly American themed attraction on first impression so it will attract Asian tourists from all over the region, but if you look deeper (and with a sensitivity to Feng Shui) it has some uniquely local characteristics that most colonial architecture is missing. It may even end up being a tourism attraction that is truly integrative of Western and Eastern symbolism for Hong Kong for some markets.

Conclusion

Can Hong Kong continue to build its new myth of itself as Asia's foremost Chinese cosmopolitan and international city and still show international tourists some challenging reminders of its colonial past? In relation to the issues outlined earlier, it is likely that it is not ready to do this, that is, if the tourism products currently offered and proposed are any indication. There appear to be three main trends. The first is for government authorities to avoid any mention of certain symbols of colonial rule in signage, tours or heritage trails, as they are still too controversial and potent for local Chinese people. The second is for cultural authorities to downplay the role of the British colonialism in Hong Kong's history and to boost the superiority of Chinese traditions and their continuity (despite the impact of colonialism) in their exhibitions as a type of historical revisionism for schools and the local community, which are being encouraged to delve deeper into their Chinese identity. The third trend is the selection of remote and poorly managed heritage attractions over closer, more central attractions (that include colonial heritage) to showcase Chinese heritage and implement poorly conceived postcolonial government policies of decentralization of tourism spending.

All these trends in policy making and presentation of heritage assets are likely to come into direct conflict with the marketing messages of HKTB promoting the integration of East/West elements and Hong Kong's cosmopolitanism. If Hong Kong was truly multicultural and cosmopolitan, symbols of colonialism could be put in context and revived, conserved or laid to rest without much alarm as has happened in Singapore and to a lesser extent in Macau. However, it is too soon after the Handover for this to occur in Hong Kong. It is also too important at this time for its emerging national myth that Hong Kong is the most superior Chinese city economically and culturally in the region for it to show what it considers to be the underside of colonial rule to visitors or the local community.

References

Abbas, A. (1997) *Hong Kong: Culture and the Politics of Disappearance*, Hong Kong: Hong Kong University Press.

Antiquities and Monuments Office (AMO) (1998) *Archaeological Discoveries from Tung Wan Tsai North, Ma Wan: One of the Ten Most Important New Archaeological Discoveries in China, 1997*, pamphlet published by the Antiquities and Monuments Office, Hong Kong SAR.

Ap, J. (2000) 'Hong Kong Disneyland-Community Reactions and its Impact on Tourism on Tourism in the Pearl River Delta Region', unpublished Paper presented at the Leisure and Entertainment Asia Conference 2000, Hong Kong SAR.

Barker, C. 2000 *Cultural Studies Theory and Practice*, London: Sage Publications.

Cheung, Sydney. C. H. (1999) 'The meanings of a heritage trail in Hong Kong', *Annals of Tourism Research* 26, 3: 570–88.

Cohen, E. (1972) 'Toward a sociology of international tourism', *Social Research* 39: 164–82.

Culture and Heritage Commission (2002) Culture and Heritage Commission. www.chc.org.hk.

du Cros, H. (2002a) Paper presented at Eighth Annual Conference of the Asia Pacific Tourism Association: Tourism Development in the Asia Pacific, Dalian, China, July.

du Cros, H. (2002b) 'Conflicting perspectives on marketing Hong Kong's cultural heritage tourism attractions', in *Strategies for the World's Cultural Heritage: Preservation in a Globalised World: Principles, Practices and Perspectives*, Madrid, 13th General Assembly of ICOMOS, pp. 319–21.

Hase, P. and Sinn, E. (1995) *Beyond the Metropolis: Villages in Hong Kong*, Hong Kong: Joint Publishing (HK).

Heng, S. (2002) 'History and SoHo in Hong Kong: The Non-/Dis-/Sub-/Re-appearance of History', *Hong Kong Cultural and Social Studies* 1, 2. www.hku.hk/hkcsp/ccex/ehkcss01/.

Home Affairs Department (HAD) (2001) 'Cultural consultation paper received public support – Tuesday, September 18, 2001'. www.had.gov.hk.

International Council on Monuments and Sites (ICOMOS) (2004) 'International Council on Monuments and Sites'. www.icomos.org, accessed 10 May.

Hong Kong Disneyland (2002) 'Disney imagineers provide first look at Adventureland and Tomorrowland inside Hong Kong Disneyland', News Release 16 October.

168   *Hilary du Cros*

Hong Kong Leisure and Cultural Services Department (LCSD) (2000) *Leisure and Cultural Link*, Hong Kong Leisure and Cultural Services Department, May.

Hong Kong Tourism Board (HKTB) (2002a) www.tdctrade.com./hksar/293.

Hong Kong Tourism Board (HKTB) (2002b) *Discover Hong Kong*. www.discoverhongkong.com.

Jacobs, J. M. (1996) *Edge of Empire: Postcolonialism and the City*, London: Routledge.

Lim, P. (2002) *Discovering Hong Kong's Cultural Heritage: Hong Kong and Kowloon (with 18 Guided Walks)*, Hong Kong: Oxford University Press.

Lowenthal, D. (1996) *Possessed by the Past*, New York: Simon and Schuster.

McDonald, G. F. and Alsford, S. (1989) *Museum for the Global Village: The Canadian Museum of Civilization*. Hull: Canadian Museum of Civilization.

McKercher, B., Ho, P., du Cros, H. and Chow, B. (2002) 'Activities based segmentation of the cultural tourism market', *Journal of Travel and Tourism Marketing* 12, 1: 23–46.

Morris, J. (1997) *Hong Kong. Epilogue to an Empire*, New York: Vintage Books.

Prideaux, B. and Kininmont, L. (1999) 'Tourism and heritage are not strangers: A study of opportunities for rural heritage museums to maximise tourism visitation', *Journal of Travel Research* 37: 299–303.

Ritzer, G. and Liska, A. (1997) '"McDisneyization" and "post-tourism": complementary perspectives on contemporary tourism', in C. Rojek and J. Urry (eds) *Touring Cultures: Transformations of Travel and Theory*, London: Routledge, pp. 96–111.

South China Morning Post (2002a) 3 March.

South China Morning Post (2002b) 22 July.

South China Morning Post (2002c) 28 October.

South China Morning Post (2002d) 31 October.

South China Morning Post (2002e) 16 November.

South China Morning Post (2002f) 20 November.

Tighe, A. J. (1986) 'The arts/tourism partnership', *Journal of Travel Research*, 24, 3: 2–5.

Zeppel, H. and Hall, C. M. (1991) 'Selling art and history: cultural heritage and tourism', *Journal of Tourism Studies* 2, 1: 29–45.

11 Globalisation and neocolonialist tourism

Reiner Jaakson

Introduction

The era of European colonialism came to an end during the 1950s and 1960s when colonies became politically independent states. In many former colonies, colonial legacies continued as political sovereignty was accompanied by economic dependence. This was the case especially in Caribbean small island states and countries in sub-Saharan Africa; former colonies in Asia fared better. Colonialism had many forms and postcolonialism too has varied widely (Thomas 1994). Colonialism and the colonial legacy have been studied extensively and there is a rich literature (Abernethy 2000; see Chapter 1). Early postcolonial nations lacked economic independence, much as today also the power of the state is being eroded by a neoliberal ideology, the rapid international movement of footloose capital, concentration of corporate wealth, privatisation of public services, outsourcing of manufacturing and services, and deregulation of trade. In the discussion that follows, the bias is to tourism from developed to developing countries, in a context of a postcolonial legacy of domination and subjugation (Chung 1993). We will argue that globalisation has changed fundamentally the nature of tourism.

An early treatise on tourism from a postcolonial perspective was by de Kadt (1979). Postcolonial metropolitan hegemony and tourism have been discussed by Britton (1980, 1982), Leheny (1995), Hiller (1976) and Davis (1978). Dependency (Wallerstein 1979), core–periphery (Malecki 1997), hinterland (Haggett 1975), development pole (Perroux 1988) and various regional development theories (Myrdal 1965; Lloyd and Dicken 1972), figure prominently in the postcolonial tourism debate, although the label postcolonial is not always used explicitly. Postcolonial tourism has been given various descriptive and emotive labels: plantation-like agricultural system (Butler 1993, Hall 1994, Beckford 1972); servility and inferiority (Husbands 1983); postcolonial identity (Palmer 1994); developing countries as subsystems (Hills and Lundgren 1977); perpetuation of a colonial space-economy (Weaver 1988); colonialism and tourism as relatives (Bruner 1989); enclave development (Freitag 1994); tourism as a sugar crop (Finney and Watson 1975); cultural imperialism (Shivji 1975); playground culture, white intrusion and fantasy (Mathews 1978); and tourism as 'new slavery' (Kinkaid 1988). The debate over whether tourism has, on balance, negative or positive effects, especially in the case of tourism from developed to developing countries, is still ongoing (Crick 1989). Tourism benefits often are regionally and socially unevenly distributed (Harrison 1995) and unemployment remains high, even in developing countries with a strong tourism sector (WTO 1998).

A polarised world of rich and poor

In 1992 the United Nations Human Development Report (United Nations Development Program 1992) showed on its cover a logo of a champagne glass with a broad cup and a narrow stem. The champagne glass has become a metaphor of a lopsided world with a huge gap between the rich and the poor. The bowl of the champagne glass represents the approximately 20 per cent of people in the world who in 1992 had 83 per cent of the world's income – 60 times the income of the poorest 20 per cent of the population. The thin stem of the champagne glass represents the poorest 20 per cent of the population who in 1992 survived on less than 2 per cent of the world's income. Tourism may well increase the overall welfare in a country, but the direct benefits from tourism often accrue to only a small percentage of the population. Elites in developing countries consistently gain disproportionately more from tourism development, as for example through soaring land values and from favouritism in the participation in investments from abroad (M. Smith 1997). If lucky, the poor may enjoy some of the indirect benefits of tourism development. 'On average, up to half of all tourism income in developing countries 'leaks' out of the destination, with much of it going to industrial nations through foreign ownership of hotels and tour companies' (Worldwatch Institute 2003: 50). Income may increase overall in a country, yet the wealth gap may remain the same or may even widen. Continued inequality amidst increasing wealth can create a frustration-aspiration syndrome where people aspire to a better life yet are unable to fulfil their heightened aspirations. In justice theory, this is the familiar trade-off between total welfare and equality (Rawls 1971).

In a world divided between rich and poor, tourism is a powerful symbol of wealth and privilege, which explains in part why in numerous countries in recent years tourists have become targets of terrorism. We see this disturbing trend to tourism targeted violence as a component of neo-colonial tourism. With global travel there is also a rising concern over the rapid spread of contagious disease, as was illustrated by the deaths from SARS (Sudden Acute Respiratory Syndrome) in several countries in 2003. 'The next plague is only a plane ride away' (as a vernacular expression puts it). The Champagne Glass World has not delivered many of the gains expected from de-colonisation. Regardless of whether the benefits of tourism outweigh the costs, of how environmentally responsible, socially conscious and culturally aware individual tourists may be, international travel from developed to developing countries is inescapably contexted in a Champagne Glass World. Since 1992, the wealth gap in the world has widened, both between and within countries, including the richest country in the world, the United States. 'American society is divided in a way that it has never been before' (Schwartz 1996: 38). In the United States, over 30 per cent of the income goes to the richest 10 per cent of the population and the poorest 10 per cent of people receive only 1.8 per cent of the income (World Bank 2002). The 2003 United Nations Human Development Report indicates that in the decade of the 1990s the standard of living declined in most of Africa and large parts of the Middle East, Latin America and Eastern Europe and 1 per cent of the world's population receives more income than the bottom 50 per cent of the population (United Nations Development Program 2003). 'Between 1980 and the late 1990s, inequality (also) increased within 48 of 73 countries' (Cornia and Court 2001: 7). In the period 1990–2001, per capita income declined in 47 countries (United Nations Development Program 2003). Trade and investment liberalisation do not reliably result in growth and economic prosperity, but they persistently perpetuate the wealth gap (Rodrik 1997).

Globalisation

Globalisation in the sense of activities that take place on a global scale is not new. Colonialism, imperialism, slavery and world wars all took place on a global scale. What distinguishes globalisation today from these historical global events is the speed and intensity of the movement of capital, labour and technology throughout the world. To Harvey (1996) a main characteristic of globalisation is the rapid movement of capital in a world where time and space have been compressed. Neoliberal capitalism has a 'penchant for achieving uneven sectoral and geographic development so as to force a divisive competitiveness between places defined at different scales' (Harvey 1996: 42). Stiglitz defines globalisation as:

> the closer integration of the countries and peoples of the world . . . brought about by the enormous reduction of costs of transportation and communication, and the breaking down of artificial barriers of the flows of goods, services, capital, knowledge, and (to a lesser extent) people across borders.
>
> (2003: 9)

Capitalism in its quest for competitiveness seeks out and exploits differences in the cost of labour, favourable trade and permissive social and environmental regulations. The result is a global 'race to the bottom' between developing countries competing to be the recipients of outsourcing from developed countries of manufacturing and service provision. Corporate globalisation creates externalities, such as environmental degradation, pollution and resource depletion, which poor countries, already stretched fiscally, are unable to cope with. The internal contradiction of hyperactive capitalism is that generating profits from the exploitation of global differences eventually erases the very differences needed to generate profits.

Jameson (1998: 60) argues that globalisation consists of twin positions of cultural heterogeneity and economic homogeneity, intertwined in a process of 'the becoming cultural of the economic, and the becoming economic of the cultural'. On the cultural front, globalisation involves a 'celebration of difference and differentiation; suddenly all the cultures around the world are placed in tolerant contact with each other in a kind of immense cultural pluralism which would be very difficult not to welcome' (pp. 56–7). Jameson (1998) also believes that globalisation has imposed a world system of standardisation and forced integration that is impossible to avoid participating in. We argue here that with standardisation and forced integration, differences are being squeezed – homogenised – out. We celebrate cultural heterogeneity yet also question how tolerant members of some cultures are of others. What should be celebrated is not so much cultural heterogeneity *per se*, but the ability of people in some communities to have been able to retain a modicum of autonomy in the face of global onslaughts. Significantly, it is precisely these, frequently remote, parts of the world, where, so far, globalisation has been resisted, that have become favourite destinations for alternative and adventure tourism.

Here we expand on the definition of globalisation as the finance led rapid expansion of corporate power. Our premise is that globalisation now affects most aspects of life, everywhere, in a neocolonial world of hypercapital. In our expanded definition of globalisation, we identify five components, which are causally linked to each other:

1   *Economic globalisation*: the rapid movement of capital (hypercapital) seeking difference as a means to sustain competitiveness and profits.
2   *Cultural globalisation*: consumerism as the predominant world cosmology.
3   *Environmental globalisation*: global impacts on the environment, particularly the atmosphere, and climate change and sea level rise.
4   *Military globalisation*: a geopolitics where the United States is the single world superpower.
5   *Militancy globalisation*: protest movements and, in many countries, increasing use of violence and terrorist acts.

Although the five components are interrelated, economic globalisation is the catalyst driving force of the entire model.

Why are we stuck with a highly unequal Champagne Glass World, almost 50 years after the end of colonialism? The list of reasons is long but includes colonial legacies of entrenched class interests; corruption and poor governance; political infighting; single sector economies based on agriculture or a natural resource; lack of infrastructure; lack of human capital. Most development theories (which we listed at the beginning), including those for tourism, have a bias that developing countries should follow a development model that originates in the developed countries in the West (Rostow 1967; Peet 1999). The distinctive feature of colonialism was 'the persistent effort of Europeans to undermine and reshape the modes of production, social institutions, cultural patterns, and value systems of indigenous peoples' (Abernethy 2000: 10). Today, Western neocolonial models of development are imposed on developing countries by owners of global capital, emboldened and empowered by international agreements on trade and investment that protect the interests of developed countries and their corporations. Global market restructuring facilitates capital that is concentrated in ownership and decentralised in application (Harrison 1995). 'Of the 100 largest economies in the world, 51 today are corporations, not countries' (Brecher *et al.* 2000: 8). Deregulation of trade opens up unrestricted access to markets in developing countries, and favours Western developed countries by allowing them to maintain import restrictions, such as on agricultural products from developing countries. 'A legacy of colonial rule in many currently independent states is a high level of vulnerability to externally generated economic and technological changes' (Abernethy 2000: 15).

Evolution of neocolonialist tourism

The mass tourism that emerged in the decades following the Second World War was pre-eminently modern, founded on legislated mandatory vacations for all, faith in a continuously growing economy, rising disposable personal income and a rationally managed travel and hospitality industry. Tourism in the 1950s coincided roughly with the beginning of decolonisation and, starting in the 1960s, with postcolonialism. A century earlier, in 1841, a pivotal event in the emergence of early modern tourism took place in England when Thomas Cook organised an all-inclusive package tour from Leicester on the Midland Counties Railway for some 570 members of the Temperance Society. Soon international tours were being organised by Cook and others, spawning a flourishing publishing industry of travel guidebooks by Murray, Baedeker and Michelin. Cook tours were the genesis of modern travel: managed, controlled and relatively cheap. But not all Cook tours were cheap and for the working or middle classes since there were also expensive tours to Egypt and other exotic locations, as prime examples of colonialist tourism during the era of railways and ocean passenger liners. Modern tourism promoted a belief in a tourist entitlement where travel was no longer seen as a privilege but as a reward of the good life for people living in rich countries and for the rich living in poor countries. With fast and relatively cheap air travel in the mid-twentieth century, mass tourism emerged as the dominant cultural motif of modern tourism.

Mass tourism was accompanied by mass problems. In recent years, this backlash has spawned a new tourism terminology aimed at a dissociation from mass tourism, by using emotive phraseology such as green tourism, responsible tourism and alternative tourism. (Mowforth and Munt 1998). The backlash has given rise to notions of 'end of tourism' (Urry 1995) and post-Fordist consumption (Lash and Urry 1994). But modernistic, Fordist tourism consumption continues to thrive, for example, in all-inclusive resort vacations, tightly monitored and controlled cruise ship vacations, and sundry sightseeing group package tours, all of which would make Henry Ford proud of how efficiently and profitably they are managed in the best assembly line tradition. But today the Champagne Glass World creates a fundamental dilemma for the *image* of tourism, of what the tourist represents, travelling in a world divided between the rich and the poor, where the poor want what the rich have, but cannot have it. As consumers of global culture, tourists are agents of globalisation. Not surprisingly, therefore, the term 'tourist' itself has acquired negative connotations surpassing even those associated with mass tourism. The new negativity is the genesis of the substitution of 'traveller' for the term 'tourist'. The traveller is portrayed as a kind of non-tourist, someone who is aware of the negative impacts of tourism and the social stigmatisation attached to tourism, especially mass tourism. It is politically incorrect to be just-a-tourist; one has to aspire to the higher office of traveller-with-a-conscience. The unsavoury image of mass tourism has evolved to a wholesale denial of the identity of the tourist as a tourist. By asserting the identity of traveller, the tourist lays claim to a superior form of tourism. The tourism-denying tourist professes to be culturally aware, socially responsible and environmentally harmless, and thereby obtains comfort from the imagined status of being a traveller, not a tourist. If only it were so simple that the mere label of traveller would absolve tourism of its problems! The ideals of tourism-denial tourism are not to be found in tourist guidebooks, but instead in books on geographic exploration and travel in extreme conditions to harsh environments. In mass marketing chain book stores in Canada, travel guides and books on exploration and travel writing are located in non-adjacent parts of the store. Chatwin, Thesinger and Theroux portray the ideals of the tourism-denying tourist, relegating tourist travel guides by Fodor, Fromer, Insight, Eyewitness, etc. to the world of mere ordinary tourists. Attempts to identify as a traveller-not-tourist may explain why eco-tourism and various forms of alternative and adventure tourisms have become popular. The desire to dissociate from tourism and to be seen instead as a traveller, remains strong. An advertisement in a Toronto daily newspaper promoted a guided tour to Turkey with the prominent heading that this tour was 'For travellers, not tourists!' With prices starting at Can$5,100, the select 18 participants (the advertised group size) may perhaps be forgiven if they indulge in make-believe that they are travellers, not tourists. An element of play such as make-believe has always been part of tourism but in neocolonialist tourism the make-believe has become the wholesale redefining of who a tourist is.

Contrast and disappearing difference

To Cohen (1974) tourism involves the seeking of novelty and change while for MacCannell (1973) tourism is a quest for authenticity. Based on these earlier views, we theorise that most tourists seek some degree of contrast and difference between home and travel destination; that the variables which describe home and destination are inversely related; and that this inverse relationship defines the degree of touristicity in tourism. However, since there are many tourisms and types of tourists, this touristicity contrast model does not apply equally to all types of tourists. The contrast between home and destination can vary according to climate, built and natural environment, culture (language, religion, food, dress, customs), recreation activities and leisure, and the perception of time. Some degree of contrast (between home and destination) will always remain, but with the spread of global culture differences continue to disappear. As consumers of global culture, tourists from all countries tend to look more and more alike. Contrast and difference are strong themes in tourism marketing and advertising, and where contrast is lacking it can be created as simulacra of the real and as staged authenticity (Cohen 1979, MacCannell 1973, 1976).

We see neocolonialist tourism as consisting of three broad categories. First, *contrast seeker* tourists for whom erosion of difference affects negatively their touristic experiences, and who seek aggressively destinations and experiences which provide a high degree of contrast. Second, *contrast indifferent* tourists who are oblivious, or who do not particularly care one way or the other, about signs of global culture. Third, *contrast avoider* tourists who seek the comforts and reminders of home, and who find solace in familiar signs of global culture. However, tourists do not necessarily consistently fit exclusively into one category only; goals may vary from one travel experience to another and even during the same trip each phase may involve its own melange of touristicity contrast. With increased travel experience, a tourist may evolve from the contrast indifferent to the contrast seeker category. Conversely, with old age or with growing concern over safety and security, preference in travel may evolve in the direction of the contrast avoider category. As contrast and differences are diluted, the surviving differences become progressively more important and are sought out, especially by the prestige conscious 'curriculum vitae builder' tourists (Panos Institute 1995). Tourists from the West to the Rest (Hall 1992) will be seen by the Rest as missionaries of the West, as a neocolonial replay of the legions of white European colonisers who roamed across the globe in past centuries.

Global culture may be comforting for the contrast avoider tourists but for the contrast seeker tourists it is experientially unexciting. Beachfront resorts, shopping arcades of chain stores, international airports, urban waterfront developments, hotels and restaurants are all essentially replications of each other, part of an international, corporate hotel and hospitality industry, and a standardised architecture and urban planning. This is global culture masquerading as local culture. Attempts to create difference – with theme parks, waterfront redevelopments, landmark public buildings designed by celebrity architects – become but further reminders of global culture and of the futility of repetitive attempts at differentiation. Cultural change works as much from the 'outside' as it does from the 'inside', as the global working through the local or glocalisation (Ruigrok and van Tulder 1995). For English native speakers, and for the large global population that is conversant in English as a second language, the tourist will be greeted in English as the *lingua franca* of global culture almost anywhere. Tourist destinations become a blend of the local and the global, where the globally familiar increasingly overshadows what has remained of the local novel. Hotels with air conditioning, clean sheets, potable water, CNN on the television and with 'food like at home', including fast-food, can be found in all large cities and in most other places as well; this is familiarity of home, everywhere away from home. The Golden Arches corporate logo of the McDonald's fast-food restaurant chain is a logo also of global culture. Tourists share experiences with other tourists who, whatever their language and nationality, are reminders of global culture since they all are dressed in familiar mass marketing clothes and carry the same electronic gadgetry. By seeking contrast, tourists by their presence erase the contrast they seek, a zero sum game.

Fear, risk and uncertainty

Risk and uncertainty have always been part of tourism. Accounts by travellers on the Grand Tour are rife with experiences of assaults and robberies, unsafe accommodation, treacherous interpreters and guides, dishonest currency exchangers and accident-prone roads and bridges due to poor maintenance (Hudson 1993). Travel was *travail* in the true sense of the French word for work; fear was part of tourism. Grand Tour travellers often had to hide their national identity while in countries which were in a state of conflict with their own country. The Cook tours of the nineteenth century (Withey 1997) were the first modern attempts to take the uncertainty out of tourism. The *travail* of travel was replaced by travel as recreation. Since the 1950s, travel by air has become efficient and affordable, although it is not necessarily always enjoyable and can sometimes be dangerous. While some degree of risk has always been part of tourism, what distinguishes neocolonialist tourism today is the sinister form of risk due to the potential of random acts of terrorism. The geopolitics of travel has regressed to a stage where one's skin colour, ethnicity and outward signs of religion determine the level of hostility and suspicion with which one is scrutinised at airports. Moreover, in addition to the risk of tourist-directed violence, there is growing fear of the spread of communicable infectious disease. During the past fifty years of modern tourism, travel was affordable and relatively safe; uncertainty was suppressed, although it was never entirely absent. During the past decade or two, in a Champagne Glass World where tourists may not always be welcomed and often are resented, fear and concern over personal safety have become part of neocolonialist tourism.

In a Champagne Glass World of inequality, tourism *qua* tourism – the concept of tourism itself – has been politicised. When terrorists attack tourists, they do so not because of grievances against tourists as individuals but because of what the tourists represent, and because of the attention terrorism is sure to generate in the global mass media. Tourists have become political pawns. The minute tourists leave home, to travel across the country or across the world, they are actors on the Champagne Glass World stage, and become potential targets. One of the first tourists killed by terrorists was in 1985 when an elderly Jewish-American passenger was singled out and murdered on the cruise ship *Achille Lauro* in the Mediterranean Sea. Tourist targeted terrorism since then has taken place in numerous countries, with increasing frequency and fury. Randomness and uncertainty are stowaways in neocolonial travels. Modernity, with its hierarchical, linear and rational organisation was able to reduce risk, whereas in contemporary network society (Castells 1996) risk is dispersed and much more difficult to control. Since the end of the Cold War, the relative military power of states has shifted dramatically and power now is concentrated in the United States as the single global superpower – a geopolitical condition that has not existed quite to that degree since the time of the Roman Empire. The United States accounts for 36 per cent of all military spending in the world (Skons 2002). After the terrorist attacks on the World Trade Center in New York in 2001, the United States embarked on a course of unilateral global militarism as part of its misguided effort to seek homeland security. However, in a Champagne Glass World conflict exists not simply between the West and the Rest, but more significantly between the rich and the poor everywhere, as part of protest movements opposed to globalisation in both the West and the Rest. Peace and security will remain elusive as long as a divided Champagne Glass World continues.

Neocolonialist tourism in a Champagne Glass World is nowhere more evident than in cruise ship tourism. Cruise ships are hedonistic floating pleasure palaces where the well-to-do from developing countries enjoy themselves in pampered luxury. With a wide range of prices, cruise ship tourism offers excellent value for the money paid, as the high numbers of repeat users indicates. Globalisation of cruise ship tourism is evident in the great concentration of the industry into three corporations that together control more than 80 per cent of the market. Most cruise ships are registered in developing countries where there are few or no regulations, whereby corporations are able to further maximise profits by being able to avoid paying their workers health and other benefits, which would have been required if the ships were registered in the United States or another developed country. Cruise ships that operate in the Caribbean do not hire from the region because workers there tend to be unionised and demand relatively high wages (Wood 2000). Instead, unskilled workers are hired from developing countries, such as Indonesia and the Philippines, and more recently from Eastern Europe. By contrast, officers and skilled workers tend to be hired mostly from developed countries. A cruise ship is the ultimate example of neocolonialist tourism in a Champagne Glass World, where the West and the Rest come into direct contact as passengers from developed countries are served by workers from developing countries. Passengers, who are overwhelmingly white, are served by mostly non-white workers. Even during a shore visit (one of the events offered during most cruises) the boundary between the West and the Rest remains intact. Shore visits typically are to a port where cruise ship tourists rarely venture beyond a familiar tourist bubble of souvenir shops and restaurants (Jaakson 2004), or to an exclusive enclave on shore owned by the cruise ship company, where passengers are entertained in a mini-resort intentionally designed to be isolated from the surrounding community.

Visualise a cruise ship that has arrived in a port on a small Caribbean island that is heavily developed with hotels and resorts. The ship is spectacularly large, one of the recently built gargantuan vessels accommodating several thousand passengers. As viewed from shore, the cruise ship is gleaming white and larger than any built structure on the island. All the passengers are white. To an unemployed black West Indian man or woman on shore, gazing at the white cruise ship of white tourists, the imagery and associations generated by the vision of the cruise ship may not be very different from what it would have been like some 150 years ago when gazing at a large sailing ship recently arrived at the colony from the mother country, with flags flying, cannons gleaming in the tropical sunshine, and officers on board pompously costumed in imperial finery.

In the 1994 Carnival King calypso competition in St Lucia, the West Indies, the winning song was 'Alien', written by Rohan Seon and sung by Mighty Pep. Calypso music has always been a vehicle for protest and social change in the Caribbean. 'Alien' became immensely popular. The song could well serve as an anthem of neocolonial tourism commenting, as it did, on feeling 'like a stranger' in one's own land.

Conclusion

Is globalisation unchallengeable? As immense as the influence is of global capital and the International Monetary Fund (IMF), World Trade Organization and World Bank, there is a growing network of non-governmental organisations (NGOs) and international protest groups and alliances opposed to globalisation. Strategies to oppose globalisation, including alternative models of economic development, have been formulated by the International Forum on Globalization (2002). Opposition to globalisation is creating common interests that transcend national and interest-group boundaries, in a movement that Brecher *et al.* (2000) call globalisation from below. Globalisation from below is not aligned along divisions of North/South, First World/Third World or developed/developing countries, because the gap between rich and poor cuts across all these divisions – a division not merely between the West and the Rest, but between the Rich and the Rest everywhere. If opposition to globalisation is to be successful, it has to take place on many different scales, and resistance strategies must not be restricted to either-or options such as resuscitating the power of the state or creating some single, global super-organisation. For example, if a Tobin tax (a tax on international financial transactions, the revenues from which could be used to ameliorate the negative impacts of globalisation) were to be successful, it would require global enforcement, as well as a stronger state.

What role is there for tourism in globalisation from below to eliminate poverty and to empower people to have a greater say in their future? While tourism can be an important income generator, on its own it cannot close the global gap between the Rich and Rest. Quite the contrary, neo-colonialist tourism often perpetuates and even exacerbates the void between the Rich and the Rest. Control of global capital invested in tourism calls for enforceable codes of acceptable conduct. For example, the illegal trade in diamonds, which is often linked to armed conflict, is being curbed by consumer awareness campaigns and by 'naming and shaming' corporations that are involved in the diamond trade. Tourism corporate codes of conduct would be difficult to enforce, although the genesis for an administrative structure may already exist in tourism NGOs, such as the Ecumenical Coalition on Third World Tourism (ECTWT). It is unlikely that, on their own, state organisations in developing countries would have the political power or wherewithal for enforcement, given their debt peonage and austerity programs imposed by the IMF. It is similarly unlikely that, on its own, some new global overseeing organisation would have sufficient power to enforce compliance. The option we are left with is action by networks of organisations, operating at different scales, and at different phases of tourism and development. There is an unrealised political power in marketplace civil disobedience and boycotts by the hundreds of millions of tourists who travel each year. The power of boycotts and protest movements opposed to globalisation has been well documented by Kline (2002). While we would not support a total boycott of travel to a poor country that desperately needs tourism income, selective tourism boycotts do form a potentially powerful means of protest. For example, child prostitution to serve tourist clients is a problem in several developing countries that have received large investments of global capital for tourism development. To eradicate child prostitution, local law enforcement and attempts to prosecute tourists in their home countries for sex crimes committed abroad, have largely failed. A boycott of tourist travel to a country where child prostitution continues unabated is an example of globalisation from below. We propose a pro-active approach, where tourists as global consumers express their protest by making selective choices of where *not* to travel. There is a huge reservoir of untapped political power in an activist tourism, waiting to be implemented for the betterment of the world.

References

Abernethy, D. (2000) *The Dynamics of Global Dominance: European Overseas Empires, 1414–1980*, New Haven, CT: Yale University Press.

Beckford, G. (1972) *Persistent Poverty: Underdevelopment in Plantation Economies of the Third World*, New York: Oxford University Press.

Brecher, J., Costello, T. and Smith, B. (2000) *Globalisation from Below: The Power of Solidarity,* Cambridge, MA: South End Press.

Britton, S. G. (1980) 'The evolution of colonial space economy', *Journal of Historical Geography* 6, 3: 251–74.

Britton, S. (1982) 'The political economy of tourism in the Third World', *Annals of Tourism Research* 9, 3: 331–58.

Bruner, E. (1989) 'Of cannibals, tourists, and ethnographers', *Cultural Anthropology* 4, 4: 438–45.

Butler, R. W. (1993) 'Tourism development in small islands: Past influence and future directions', in D. Lockhart, D. Drakakis-Smith and J. Schembri (eds) *The Development Process in Small Island States*, Case Studies London: Pinter, pp. 11–49.

Castells, M. (1996) *The Rise of Network Society*, Oxford: Blackwell.

Chung, H. (1993) 'People's spirituality and tourism', *Contours* 6, 7/8: 19–24.

Cohen, E. (1974) 'Who is a tourist? A conceptual clarification', *Sociological Review* 22: 527–55.

Cohen, E. (1979) 'Rethinking the sociology of tourism', *Annals of Tourism Research* 16, 1: 30–61

Cornia, G. A. and Court, J. (2001) *Inequality, Growth and Poverty in the Era of Liberalization and Globalization*, Helsinki, Finland: World Institute for Development Economics Research.

Crick, M. (1989) 'Representations of international tourism in the social sciences: sun, sex, sights, savings, and servility', *Annual Review of Anthropology* 18: 307–44.

Davis, D. E. (1978) 'Development and the tourist industry in Third World countries', *Society and Leisure* 1: 301–22.

de Kadt, E. (ed.) (1979) *Tourism: Passport to Development? Perspectives on the Social and Cultural Effects of Tourism in Developing Countries*, New York: Oxford University Press.

Finney, B. and Watson, A. (ed.) (1975) *A New Kind of Sugar: Tourism in the Pacific*, Honolulu: East-West Culture Learning Institute.

Freitag, T. (1994) 'Enclave tourism development: for whom the benefits roll?', *Annals of Tourism Research* 21, 3: 538–54.

Haggett, P. (1975) *Geography: A Modern Synthesis,* New York: Harper and Row.

Hall, C. M. (1994) 'Is tourism still the plantation economy of the South Pacific? The case of Fiji', *Tourism Recreation Research* 19, 1: 41–8.

Hall, S. (1992) 'The West and the Rest: discourse and power', in S. Hall and B. Gieben (eds) *Formations of Modernity*, Oxford: Polity Press.

Harrison, D. (1995) 'International tourism and the less developed countries: A background', in D. Harrison (ed.) *Tourism and the Less Developed Countries*, Chichester: John Wiley and Sons, pp. 1–18.

Harvey, D. (1996) *Justice, Nature and the Geography of Difference*, Oxford: Blackwell.

Hiller, H. L. (1976) 'Escapism, penetration and response: industrial tourism and the Caribbean', *Caribbean Studies* 16, 2: 92–116.

Hills, T. L. and Lundgren, J. (1977) 'The impact of tourism in the Caribbean', *Annals of Tourism Research* 4, 5: 248–67.

Hudson, R. (ed.) (1993) *The Grand Tour 1592–1796*, London: The Folio Society.

Husbands, W. (1983) 'The genesis of tourism in Barbados: further on the welcoming society', *Caribbean Geographer* 2: 107–20

International Forum on Globalization (2002) *Alternatives to Economic Globalization*, San Francisco: Berrett-Koehler.

Jaakson, R. (2004) 'Beyond the tourist bubble? Cruise ship passengers in port', *Annals of Tourism Research*, in press.

Jameson, F. (1998) 'Notes on globalization as a philosophical issue', in F. Jameson and M. Miyoshi (eds) *The Cultures of Globalization*, Durham, NC: Duke University Press, pp. 54–77.

Kincaid, J. (1988) *A Small Place*, London: Virago.

Kline, N. (2002) *Fences and Windows: Dispatches from the Front lines of the Globalization Debate*, Toronto: Vintage Canada.

Lash, S. and Urry, J. (1994) *Economies of Signs and Space*, London: Sage.

Leheny, D. (1995) 'The political economy of sex tourism', *Annals of Tourism Research* 22, 2: 367–84.

Lloyd, P. and Dicken, P. (1972) *Location in Space: A Theoretical Approach to Economic Geography*, London: Harper and Row.

MacCannell, D. (1973) 'Staged authenticity: arrangements of social space in tourist settings', *American Journal of Sociology* 79, 3: 589–603.

MacCannell, D. (1976) *The Tourist: A New Theory of the Leisure Class*, New York: Sulouken.

Malecki, E. (1997) *Technology and Economic Development* (2nd edn) Harlow: Longman.

Mathews, H. (1978) *International Tourism: A Political and Social Analysis*, Cambridge, MA: Schenkman.

Mowforth, M. and Munt, I. (1998) *Tourism and Sustainability: New Tourism in the Third World*, London: Routledge.

Myrdal, G. (1965) *Economic Theory and Under-Developed Regions*, London: University Paperbacks.

Palmer, C. (1994) 'Tourism and colonialism: The experience of the Bahamas', *Annals of Tourism Research* 21, 4: 792–811.

Panos Institute (1995) 'Ecotourism: paradise gained or paradise lost?', *Panos media briefing* 14, London: Panos Institute.

Peet, R. (1999) *Theories of Development*, New York: Guilford Press.

Perroux, F. (1988) 'The pole of development's new place in a general theory of economic activity', in. B. Higgins and D. Savoie (eds) *Regional Economic Development: Essays in Honour of Francois Perroux*, Boston, MA: Hyman, pp. 48–76.

Rawls, J. (1971) *A Theory of Justice*, Cambridge, MA: Harvard University Press.

Rodrik, D. (1997) *Has Globalization Gone Too Far?*, Washington, DC: Institute for International Economics.

Rostow, W. (1967) *Stages of Economic Growth: A Non-Communist Manifesto*, Cambridge: Cambridge University Press.

Ruigrok, W. and van Tulder, R. (1995) *The Logic of International Restructuring*, London: Routledge.

Schwartz, B. (1996) 'Reflections on inequality: the promise of American life', *World Policy Journal* (winter 1995–6).

Shivji, I. (1975) *Tourism and Socialist Development*, Dar Es Salam: Tanzania Publishing House.

Skons, E. (2002) *Military Expenditure*, Stockholm: Stockholm International Peace Research Institute.

Smith, M.E. (1997) 'Hegemony and elite capital: the tools of tourism', in E. Chambers (ed.) *Tourism and Culture: An Applied Perspective*, Albany: State University of New York Press, pp. 199–214.

Stiglitz, J. (2003) *Globalization and its Discontents*. London: W.W. Norton.

Thomas, N. (1994) *Colonialism's Culture: Anthropology, Travel and Government*, Oxford: Polity Press.

United Nations Development Program (1992) *Human Development Report 1992*, New York: Oxford University Press.

United Nations Development Program (2003) *Human Development Report 2003*, New York: Oxford University Press.

Urry, J. (1995) *Consuming Places*, London: Routledge.

Wallerstein, I. (1979) *The Capitalist World Economy*. Cambridge: Cambridge University Press.

Weaver, D. (1988) 'The evolution of 'Plantation' tourism landscape on the Caribbean Island of Antigua', *Tijdschrift voor Economische en Sociale Geographie* 79, 5: 319–31.

Withey, L. (1997) *Grand Tours and Cook's Tours: A History of Leisure Travel, 1750 to 1915*, New York: William Morrow and Company.

Wood, R. E. (2000) 'Caribbean cruise tourism: Globalization at sea', *Annals of Tourism Research* 27, 2: 345–70.

World Bank (2002) *World Development Indicators 2002*, Washington, DC: World Bank.

Worldwatch Institute (2003) *Vital Signs 2003: The Trends that are Shaping Our Future*, New York: W.W. Norton.

WTO (1998) *Tourism 2020 Vision: Influences, Directional Flows and Key Trends*, Madrid: World Tourism Organization.

12 Conclusion

Hazel Tucker and C. Michael Hall

> Postcolonial studies questions the violence that has often accompanied cultural interaction and attempts to frame explanations of it as well as to provide alternate models of accommodation or getting along. It also proposes practical models of ending or channelling conflict, often by rethinking the nature of identity in situations where groups come together and interact.
>
> (Schwarz 2000: 5).

> Nowadays one cannot be serious or systematic about world political economy if one leaves international tourism out of the picture.
>
> (Crick 1995: 210)

These two quotations reiterate the central message of this book in that they both highlight the links between postcolonial studies and tourism studies and indicate the important contribution that each can make to the other. It has become clear throughout the chapters in this volume that, in particular, tourism studies can no longer exclude postcolonial theory and criticism in its analytical framework. This is so for two main reasons: first, contemporary tourism practice is both deeply embedded in and reinforcing of postcolonial relationships (e.g. Edensor 1998; Chapter 1, this volume); second, as the introductory chapter and Hollinshead's contribution to this volume (Chapter 2) indicate, much of the academic commentary on tourism has echoed and thus perpetuated colonial discourse. If postcolonial studies is about questioning the violence that accompanies cultural interaction and rethinking the nature of identity in situations where groups come together and interact (Schwarz 2000), therefore, then postcolonial theory needs to be given a much more central place, especially if theorisation and teaching in tourism studies is to keep up with current thinking in related disciplines such as cultural geography and anthropology. Both of these points will be further expanded upon in this concluding chapter.

Tourism relationships as an echo of colonial relationships

As Loomba (1998) has pointed out, a debate exists within postcolonial studies about the appropriateness of its name in that the term 'post' implies that the inequities of colonial rule are over. Indeed, the research upon which the various chapters in the volume are based has shown that the practice of contemporary international tourism indicates that the economic structures, cultural representations and exploitative relationships that were previously based in colonialism are far from over.

Accordingly, it might be said that neocolonial is a better term to describe many of the tourism contexts discussed in this book. Further to this, Alva (1995) has suggested that postcoloniality should refer not so much to the state of being after colonialism, as to an oppositionality to imperialising/ colonising discourses and practices. Postcolonial studies, then, should be viewed as the analysis or unpicking of, as well as the contestation of, the legacies of colonialism and colonial domination. This view fits with many of the chapters in this volume, and it also reflects at least the first three of the four interweaving areas of investigation in postcolonial studies set out by Ashcroft *et al.* (1989) and discussed earlier by Hall and Tucker in the introductory chapter to this volume: hegemony, language and text and place and displacement.

Many of the chapters here have shown how the hegemonic structures of language and representation have created particular conceptions of 'truth' and 'reality' for both tourism practice and tourism destinations. The study by Simmons has shown how colonial discourse is all-pervading in contemporary travel discourse in travel articles in Australian popular magazines. Her study highlights how it is that 'the remains of imperialism not only linger in Western imaginations about cultural, racial and gender superiority, but are in fact central to a contemporary travel discourse that is reproduced in the written travelogue text' (Simmons, this volume). Indeed, the presence of colonial discourse in travel fantasy is prominent throughout the postcolonial world. A favourite tourist activity in Singapore is to visit the famous Raffles Hotel for a cocktail (Henderson, this volume), and in Hong Kong many tourists have their photograph taken sitting in a rickshaw pretending to be pulled around by a Chinese 'pigeon-chested pensioner' (du Cros, this volume). The preservation and use of colonial buildings and other monuments as tourist attractions is also prominent in Singapore, as well as Penang in Malaysia and Levuka in Fiji, as discussed in the chapters by Henderson and Fisher respectively.

Such promotion and use of the colonial past for tourism does not go uncontested in the postcolonial setting, however. Fisher (this volume) has discussed the contestation of the preservation of colonial built heritage in Levuka by the indigenous Fijians at an individual or personal level.

Du Cros (this volume) describes the debate that took place in the Hong Kong media regarding a proposal to establish a working rickshaw stand on the southern side of Hong Kong island in order to revive this colonial practice with 'more authenticity' than the static rickshaw photograph opportunity. Du Cros explains that the fact that the target market for this attraction comprises the population of the previous colonial power, and that 'the recent and more direct experience of the inequity of colonialism and the need to empower the Chinese after many years of official and unofficial discrimination' (this volume) is likely to be at the forefront not only of the rickshaw debate but also broader issues of tourism and representation.

Other chapters in the volume, as well as that of du Cros, have discussed the contestation and negotiation of using colonial legacies in tourism at the level of postcolonial government and policy. In some tourism destinations it seems that a situation of ambivalence and even downright confusion exists for governments that are driven on the one hand to preserve and promote colonial heritage for the purposes of generating revenue from tourism, while on the other hand being careful not to dwell upon and over-glorify the colonial past in their postcolonial national myth-making. Again, Henderson's discussion of Singapore and Malaysia has provided a good example of this tension between the economic motives of tourism promotion and development and the ideological, political and social currents that inevitably underpin any heritage conservation movement.

Besides the tensions inherent in the role that heritage construction plays in national myth-making, an ambivalence is also often felt by postcolonial governments when direct economic ties are retained, or indeed created anew, with the ex-colonising nation. In the introductory chapter to this volume, the idea of tourism being a form of imperialism and also an example of a new colonial plantation economy was related to the particular core-periphery relationship that exists when air services, resort and recreational developments are owned by foreign interests. Akama's chapter on wildlife safari tourism in Kenya (Chapter 9) has provided a good example of the way that tourism policies and programmes can still be strongly influenced by exogenous factors based in the (former colonial) Western metropolis. Since the initial investment costs for the large-scale, capital intensive tourism projects are often too high for postcolonial governments and indigenous investors, they depend on external capital investment usually from multinational conglomerates. This process has also been described by Jaakson in Chapter 11 in making the point that while classical colonisation was state globalisation, postcolonialism is private-sector globalisation. Postcolonialism, Jaakson argues, is characterised by a much weaker role of the state and a greater role of multinational corporations. Jaakson shows a postcolonial optimism, however, in answering the question of whether such globalisation and neo-imperialism is unchallengeable. While Jaakson acknowledges that state organisations in developing countries are unlikely to have the political power or wherewithal to challenge this form of neo-imperialism, the chapter does hail the opportunities for the marketplace itself to enact such a challenge through boycotts and protest against selected tourism corporations and activities.

This last point raises the very important question asked by Simmons (Chapter 3) in the conclusion to her chapter, which is: how (and we might add to what extent) do present-day tourists, whether Westerners or not, negotiate, dismantle, resist or sustain the colonial elements of contemporary travel discourse and industry in their travel practices? This, along with the same question regarding all people engaged with and in particular marketing tourism is clearly a crucial topic of future research in tourism studies. However, the same question needs to be asked in relation to those researching and commentating on tourism: How and to what extent do they/we negotiate, dismantle, resist or sustain the colonial elements of contemporary travel discourse and industry? This is perhaps the most pertinent issue with which to conclude in the final section of this book.

Tourism scholarship as echo of colonial/postcolonial discourse

The chapter by Marschall in this volume (Chapter 6) has described the process whereby 'postcolonial agents' (those previously marginalised who have become empowered to 'speak') have used heritage to counter the biased accounts of the coloniser and to tell *their* side of the story. Heritage can thus be a means for the postcolonial agent to assert a new (decolonised) identity. Ambivalence arises in this postcolonial theorising, however, when it is the postcolonial agent's precolonial heritage being used to mark the end of colonisation. The cause of the ambivalence is that the precolonial heritage might tend to draw upon images and representations of the colon*ised* that were generated and used by the colon*isers*. Marschall takes a postcolonial stance on this practice by arguing that the use of heritage 'that mimics or imitates Western models can thus be interpreted as a strategy of the postcolonial agent to appropriate the (visual) language of the coloniser in order to "write back" (Ashcroft *et al.* 1989), to respond to and "de-scribe" the discourses of the coloniser' (this volume).

This reading by Marschall certainly reflects the postcolonial discourse of Bhabha as described by Hollinshead in Chapter 2: 'At times, such in-between peoples will appear to mimic or imitate the cultural institutions of their erstwhile colonisers, and at times they may even appear to parody the supposed voices of their own precolonial "Other"' (this volume). Hollinshead goes on to point out that, 'It takes a certain degree of embed-dedness in local circumstance to determine which of the newly gelling or the reconsolidating codes-of-being and intensities-of-affiliation which flare up within these kinds of in-between communities really matter' (this volume). In other words, it is often quite difficult for tourism commenta-tors to know when cultural identities which appear to mimic and play according to colonial representations should be read as empowering forms of 'cultural hybridity' and when they should be read as the result of violent acts of colonial discourse which can be 'incrementally and savagely crippling' (Hollinshead, this volume).

In the chapter by Wels in this volume (Chapter 5) it is argued that the colonial myths and fantasies that shaped the European social constructions of African landscapes and peoples continue to play a powerful role in shaping the current gaze of Europeans in postcolonial tourism today. Wels's point concerns the violence in the way that the continued colonial representation of the Bushmen of southern Africa, who are perhaps the most victimised and brutalised people in the colonial history of southern Africa, has led to the Bushmen being presented for tourism today as a beautiful people living in primeval paradise. The reason for this is that:

> That is the Otherness (i.e. 'them') Europeans (i.e. 'us') want to experience in Africa and for which they are prepared to pay money. This is the imagery or staged authenticity to which the tour-operators have to relate in their brochures in order to persuade clients/tourists to book a holiday with them. This is the imagery of African culture which cultural tourism must reflect in its programmes.
>
> (Wels, this volume)

Hence, Wels is pointing out the way in which tourism can be, as Hollinshead also warns, 'the violence-rendering rhetorical instrument of imperialism, perpetually dealing in Eurocentric accounts' (this volume). Unfortunately, the same imperialist imagery has often been maintained by commentators on tourism as well as those in the tourism industry, showing a desire by them also to fix the 'ethnic' identities of peoples in tourism destinations into perpetual 'Otherness'. As Bruner (2001: 881) has commented, 'Tourism scholarship thus aligns itself with tourism marketing, in that scholars tend to work within the frame of the commercial versions of their sites'. Yet, because it will never be possible '"to return to or to rediscover an absolute precolonial cultural purity, nor to create national or regional functions entirely independent of the historical implication in the European colonial enterprise' (Ashcroft *et al.* 1989: 195–6), then tourism commentary needs to adopt fully the postcolonial way of thinking that 'identity is a matter of "becoming" as well as of "being"' (Loomba 1998: 181). Any discussants of authenticity, tradition and cultural commodification in relation to tourism therefore need (at least to attempt) to locate and evaluate the ideological, political and aesthetic bases of their analysis. There is a need, in other words, to recognise the emergent nature of culture and identity, and to acknowledge and celebrate cultural hybridity and transnationalism rather than lamenting the loss of some *a priori* notion of cultural tradition (Tucker 2003; Coles *et al.* 2004).

Similarly, care should be taken not to be too totalising in critical analyses of the colonial elements of the ideological and political bases of representations in the tourism and heritage industries. In Chapter 2 of this volume, Hollinshead acknowledges several tourism theorists who, during the 1990s, achieved much in their writing on 'the declarative value of tourism'. Is it possible to have it both ways in postcolonial theory, however? Can we, on the one hand, say that 'the tourism industry – in close alliance with the local/regional heritage industry – frequently constitutes a scarcely stoppable collaborative force which converts local places into extremely tightly scripted destinations' (Hollinshead, this volume, drawing on Kirshenblatt-Glimblett 1998), while on the other hand, painting tourism as 'the identity business of textual negotiation, thereby helping one and all arrive at the right kind of new sense definitions for those sought or celebrated destinations' (also Hollinshead, this volume, drawing on Bhabha 1994)?

There is clearly a need for more nuanced and more ethnographically (Brunner 2001) and ethnohistorically (Duval, this volume) based analyses of the range of tourist displays within any one cultural area. Indeed, Hollinshead also pointed out in Chapter 2 that while some tourism representations will be based on colonial narratives, others will be based on resistance to those same colonial narratives. Many of the areas of tourism research that are now becoming central to the field, such as identity and representation, migration, power, sex, gender and disabilities will all benefit from the contributions that postcolonial theory and criticism can bring. It is hoped, therefore, that this book will help set a particular epistemological and ontological direction within tourism studies, in which new paths, theoretical directions for research, and discourses are opened up rather than being closed off.

References

Alva, J. J. K. de (1995) 'The Postcolonization of the (Latin) American Experience, A Reconsideration of "Colonialism", "Postcolonialism" and "Mestizaje"', in G. Prakash (ed.) *After Colonialism, Imperialism Histories and Postcolonial Displacements*, Princeton NJ: Princeton University Press, pp. 241–75.

Ashcroft, B., Griffiths, G. and Tiffin, H. (1989) *The Empire Writes Back: Theory and Practice in Postcolonial Literatures*, London: Routledge.

Bhabha, H. (1994) *The Location of Culture*, London: Routledge.

Bruner, E. (2001) 'The Maasai and the Lion King: authenticity, nationalism, and globalisation in African tourism', *American Ethnologist* 28, 4: 881–908.

Coles, T., Duval, D. and Hall, C.M. (2004) 'Tourism, mobility and global communities: new approaches to theorising tourism and tourist spaces', in W. Theobold (ed.) *Global Tourism*, Oxford: Heinemann, in press.

Crick, M. (1995) 'The anthropologist as tourist: an identity in question', in M. Lanfant, J. Allcock and E. Bruner (eds) *International Tourism: Identity and Change*, London: Sage: 205–23.

Edensor, T. (1998) *Tourists at the Taj: Performance and Meaning at a Symbolic Site*, London: Routledge.

Kirshenblatt-Gimblett, B. (1998) *Destination Culture: Tourism, Museums and Heritage*, Berkeley, CA: University of California Press.

Loomba, A. (1998) *Colonialism/Postcolonialism*, London: Routledge.

Schwarz, H. (2000) 'Mission impossible: introducing postcolonial studies in the US academy', in H. Schwarz and S. Ray (eds) *A Companion to Postcolonial Studies*, Malden, MA: Blackwell Publishers, pp. 1–20.

Tucker, H. (2003) *Living with Tourism: Negotiating Identities in a Turkish Village*, London: Routledge.

Index